Sociology Saves the Planet

Highlighting how the environment and society are intrinsically linked, this book argues that environmental concerns need to be treated as a core concept in the study of sociology.

Given its focus on inequality and the constituent elements of the social world, sociology has often been accused of negligence regarding the urgency of the world's environmental crisis. *Sociology Saves the Planet* corrects this misperception by integrating the theme of environment and society to highlight the intrinsic value a sociological perspective brings to our understanding of the current ecological crisis. The author first draws out the origins of sociology in the social and ecological transformations of the industrial revolution. In accounting for the social upheavals of the 19th century, Emile Durkheim, Karl Marx, and Max Weber all provided key insights into the changing nature of human organization and exploitation of the natural world. Second, readers will explore sociological perspectives developed since that time, grounded in evidence-based research, which highlight the inextricable connection between environment and society. Special attention is devoted to the dual role of people as producers and consumers in the modern context. Lastly, this book examines the significance of major categories of social difference regarding the current environmental crisis. In that regard the question of environmental justice is paramount, illuminating both the disproportionate benefit of natural resource exploitation to those countries and individuals with higher socioeconomic status, and the greater exposure to environmental hazard among those with less. Averting global calamity requires we recognize the unequal social impacts of the environmental crisis while valorizing inclusivity and the diversity of human experience in our search for solutions.

Designed for introductory courses, this book is essential reading for sociology students and will be of interest to students and academics studying environment and sustainability more broadly.

Thomas Macias is a Professor of Sociology at the University of Vermont, USA. He is the author of *Mestizo in America: Generations of Mexican Ethnicity in the Suburban Southwest* (2006).

Routledge Explorations in Environmental Studies

For more information about this series, please visit: www.routledge.com/ Routledge-Explorations-in-Environmental-Studies/book-series/REES

Sociology Saves the Planet

An Introduction to Socioecological
Thinking and Practice

Thomas Macias

Routledge
Taylor & Francis Group

LONDON AND NEW YORK

from Routledge

First published 2022
by Routledge
2 Park Square, Milton Park, Abingdon, Oxon OX14 4RN

and by Routledge
605 Third Avenue, New York, NY 10158

Routledge is an imprint of the Taylor & Francis Group, an informa business

British Library Cataloguing-in-Publication Data
A catalogue record for this book is available from the British Library

Library of Congress Cataloging-in-Publication Data
Names: Macias, Thomas, 1967– author.
Title: Sociology saves the planet: an introduction to socioecological
thinking and practice / Thomas Macias.
Description: Abingdon, Oxon; New York, NY: Routledge, 2022. |
Series: Routledge explorations in environmental studies | Includes
bibliographical references and index. |
Identifiers: LCCN 2021032276 (print) | LCCN 2021032277 (ebook) |
ISBN 9780367627591 (hardback) | ISBN 9780367627577 (paperback) |
ISBN 9781003110668 (ebook)
Subjects: LCSH: Environmental sociology. | Human ecology.
Classification: LCC GE195 .M324 2022 (print) |
LCC GE195 (ebook) | DDC 304.2—dc23
LC record available at https://lccn.loc.gov/2021032276
LC ebook record available at https://lccn.loc.gov/2021032277

ISBN: 978-0-367-62759-1 (hbk)
ISBN: 978-0-367-62757-7 (pbk)
ISBN: 978-1-003-11066-8 (ebk)

DOI: 10.4324/9781003110668

Typeset in Bembo
by codeMantra

For Leïla and Enzo

Contents

Acknowledgments

This book grew out of nearly two decades teaching SOC 001, An Introduction to Sociology, at the University of Vermont. I would like to recognize my undergraduate students for encouraging me to rethink how this subject could be taught with an underlying emphasis on sustainability and the nexus of environment and society. Institutional support to develop SOC 001 for university-wide sustainability credit was provided by UVM through the Sustainability Faculty Fellows Program. Additional motivation for this project can be traced back to a semi-regular interdisciplinary meeting with colleagues working at the intersection of human culture and the natural world. In that regard, I am grateful for the insights of Rachelle Gould, Mark Usher, Cheryl Morse, Adrian Ivakhiv, Tyler Doggett, Cameron Davis, and Pablo Bose. As regards research and writing, my year as a Fulbright Scholar at the IMéRA Institute for Advanced Study at the University of Aix-Marseille was indispensable. Special thanks to Raouf Boucekkine and Pascale Hurtado at the Institute for providing a stimulating and supportive atmosphere for engaging in research conversations with colleagues across the natural and social sciences and the humanities. At Routledge, I would like to thank Hannah Ferguson and John Baddeley for their editorial guidance throughout the writing process. I am also grateful for the close reading and thoughtful comments provided by four anonymous reviewers which had a significant influence on the final draft of this book. I have been fortunate to have my former student Isabella Abraham as both research assistant on the book project and teaching assistant for SOC 001. In the Macias-Aunave household, my partnership with Marielle continues to surprise; her ability to connect with others while engaging with nature – be it while tending the front yard flowerbed, on the bike trails, or during a weeklong jaunt over the bottom half of the GR20 – is an ongoing source of inspiration which, at least in part, explains my great fondness for her. Our children, Leïla and Enzo, are a treasure, a gift and boundless font of joy and meaning – like a key unlocking an iron cage out into the enchanted garden of life. I dedicate this book to them.

1 The origins of a socioecological imagination

My Peace Corps assignment in Quepos, Costa Rica was not a hardship post by any standard. There, I managed a micro-enterprise loan fund for farmers and small businesses in need of low-interest credit for seeds and machine parts. Along with immersing myself in a new social and cultural context, I entered into natural surroundings altogether different from where I grew up in Phoenix, Arizona. In that regard, my days off were a revelation. The small house where I rented a room was half a mile from the Pacific Ocean and a 25-minute bus ride to the white sand beaches of Manuel Antonio National Park. There, the Central American coast became a regular part of my life. I have vivid memories of red-tailed macaws soaring above the thick green forest, a white-faced capuchin monkey stealing my sandwich, and the saltwater glistening on my skin as I swam out to sea.

One sunny afternoon during an early attempt at body surfing, I felt myself being pulled away from shore. At first, this was a welcome sensation, saving me some effort to get near where the waves were breaking. Soon I realized, however, things were out of control. The beach appeared further and further away, and my turning around to swim toward it did little to slow the momentum of the current I had been drawn into. I am not a bad swimmer, but growing up in the Sonora Desert did not give me any advantage in this regard; my imminent drowning seemed like a real possibility. Fortunately, something clicked.

It was my high school physics teacher, I believe, who once rendered on the chalkboard a rip current to demonstrate how this particular wave motion can both benefit surfers who harness it for a free ride, and pose a lethal risk to unwary neophytes like myself. The trick for the latter group is to fight the desperate urge to swim to shore – you will only waste your energy and increase your chance of drowning. Instead, swim parallel to the coastline until you are out of the current's tow. That is what I did and, after an exhausting 20 minutes or so, I finally reached the beach. Staggering and nearly out of breath, I looked down and saw a US one-dollar bill wash up near my feet in the sand as if it to say, "Congratulations – this is what you're worth."

This made me laugh. At a moment when nature conspired to remind me of the inherent value of my own life, attaching a monetary amount to it – any amount – seemed ridiculous. The mediation of cash currency was not

DOI: 10.4324/9781003110668-1

required to comprehend either my self-worth or my respect for the natural forces that both filled me with humility and made my time in Costa Rica so exhilarating. Fortunately, we do not have to be brought face-to-face with mortality to have this realization. Much research supports the assertion that being in nature invigorates us. We are happier, healthier and less anxious when we spend more time in the natural world. Moreover, evidence is growing that those of us who spend time outdoors care more about nature and are more willing to do something to protect it than those who stay inside. The natural environment is good for us and, with increased exposure, we can be good for the environment, too.

The objective of this book is to provide an introduction to how nature fits into a broader sociological understanding of modern life; both, the impact human beings have on the global ecosystem, but also how nature has and continues to shape who we are as people in society. In some ways, sociology's relationship with nature and the natural sciences has been a bit like my relationship with the rip current that one afternoon in Costa Rica. Sociology emerged as an academic discipline near the end of the 19th century in the wake of Charles Darwin's legacy of discovery that culminated in the publication of his paradigm-shifting opus, *On the Origin of Species* in 1859. There, Darwin described the ways species adapted gradually over time to their changing habitats. In Darwin's theory of evolution, no supreme being or conscious effort was required to create the vast universe of living things found on Earth; just variations of species characteristics in the population and the tendency for some characteristics to be more advantageous in the current environment than others.

Darwin's revolutionary thinking forever changed human understanding of the natural world and, not surprisingly, affected the way we think of the human social world as well. Not long after *On the Origin of Species* was published, Herbert Spencer, another wealthy Victorian academic, coined the term "survival of the fittest" to describe his application of evolutionary theory to 19th-century England (Spencer, 1969). Just as I rode the current out to sea, Spencer devised his own "social Darwinism" to explain why vast divisions in socioeconomic class persisted despite the economic advances of the industrial revolution. The wealthy were simply better adapted to the contemporary environment of capitalist accumulation and profit in a market economy, while the poor and working-class remained at the bottom. In his day, Spencer was one of the most influential thinkers in Europe and America, reassuring those in positions of privilege of their rightful place in the natural scheme of things, while easing their minds about ever having to address the suffering and injustices faced by those in society's lower rungs.

Drawing a line between the natural and social sciences

Radical change through revolution

Fortunately, not all social thinkers were swept away by the eat-or-be-eaten thinking of social Darwinism. Modern societies were different from the natural world. Yes, they also evolved – quite rapidly, in fact, during the 19th century – but the high concentration of wealth and impoverishment of the working-class were not inevitable. Not long before, the French Revolution proved that a ruling class of monarchs and their minions could be toppled rapidly with sufficient will from those with greater numbers from below. With that sea change in the social hierarchy not too far in the past, a radical German intellectual by the name of Karl Marx began advocating for another revolution to bring down the ruling class. This time, however, it would be the capitalist owners of factories and other profit-generating properties who would be ousted by their own workers. A communist revolution would once and for all redistribute the wealth and property accumulated and concentrated under capitalism to all workers fairly and equitably – quite a different take on social evolution from that espoused by Spencer.

In thinking about humans as a species, Marx focused on our relationship to work – through our labor we derive meaning and purpose in life. In an ideal world we would all be able to find work that we enjoyed and through which we could define our existence with intention (Engels & Marx, 2004). By contrast, under capitalism, workers are primarily incentivized to labor for wages, job satisfaction being a secondary or tertiary concern. In that way, little by little, the capitalist system dehumanizes the population, with the worth of employees to their employers calculated as simply another cost in the creation of profit.

Solidarity as a source of hope and optimism

Another turn away from the dominant current of social Darwinism came from a French scholar who by the end of the 19th century was arguably the most ardent promoter of sociology as an academic discipline. It was clear to Emile Durkheim that modern society had developed in some disturbing ways from its origins in rural community settings. Notably, people in cities did not appear to be as socially connected with others as they were in small communities where everyone knew each other and shared common skills and work based in agriculture. Unlike Marx, Durkheim was not a radical willing to tear apart the social order. He did, however, see some analogies between society and the natural world which gave him hope. If people living in cities could find a way to integrate socially, they should be able to find a way to counter the disintegrating tendencies of 19th-century cities.

In the urban setting, he argued, people did not particularly trust each other and the lack of meaningful connection to others was a constant threat to civility and democratic process. By participating in voluntary or work-based organizations, Durkheim believed people living modern lives could reconnect, and society could attain a sort of social equilibrium. Inspired by symbiotic patterns in the natural world, "organic solidarity" was the term he used to describe the human ecology of profound interdependence evident in modern life where we all rely on people we will not necessarily ever meet – engineers, farmworkers, nurses, dockworkers, etc. – for our existence (Durkheim, 2014). Durkheim steered far away from social Darwinism, but he was keenly aware of the high esteem to which the natural sciences were held by the broader public. Sociology, he felt, could attain similar levels of recognition if, like plant biologists, chemists, and paleontologists, it too grounded its arguments in fact-based scientific research.

Bias and the unique advantage of understanding in sociological research

While Durkheim saw the natural sciences as an empirical model for sociology to follow, German sociologist Max Weber was much more guarded about the possibility that the social sciences could ever follow such a path. For one thing, the subjects of research – human beings – were simply different from the plants, animals, and other arrangements of molecules typically studied in the natural sciences. It was not that humans did not also constitute a particular arrangement of molecules, but when humans study other humans there is a level of understanding and empathy that does not exist between the researcher and the object of study in the natural sciences. As sociologists, we can to some degree put ourselves in the place of the people we are studying, a possibility greatly facilitated by the use of language. Unlike research on igneous rocks or hammerhead sharks, for example, we can ask other humans about their subjective experience, i.e., what a particular event, issue or stage in life actually means to them.

Weber thus brought a good deal of humility to the discipline of sociology, casting doubt on the possibility of ever attaining a purely objective understanding of social reality. We are all, even sociologists, human and though we must strive to approach something close to objectivity in our research, we also must also recognize the inherent biases we bring to bear based on our particular histories and life experience. The German name Weber gave to this form of meaningful understanding was *verstehen*, both an upfront acknowledgement of the potential for human bias in social science research, and the unique advantage available to sociologists for understanding motivation and culture among the people they study (Weber, 1978).

From a science apart to a decoupling from nature

Since its inception, then, sociology emerged from the shadow of the natural sciences which it sought to emulate in terms of its practical application of the scientific method, but also in terms of the intellectual sway of its empirically grounded arguments. Sociology thus did not develop so much in opposition to an understanding of the natural world as it was interested in creating a distinctive scholarly enterprise within which the relationship of humans with other humans was seen as necessitating the creation of a separate science of society.

Over time, however, sociology began to decouple itself from anything having to do with the natural world. By the mid-20th century, leading American sociologists such as Talcott Parsons, devised elaborate, abstract schemes of social systems and action that had little need for explaining social structure or the culture of post-war abundance in terms of its origins in nature (Parsons, 1949). When sociologists did reference "materialism," this usually signaled an affinity for Marx's *historical materialism*, an assertion that history was driven not by human ideals but rather by matter itself, lending a kind of inevitability to the natural course of the worker revolution. That was a long way from recognizing human society's place in the natural ecology. By the 1960s, however, things were changing.

The publication of Rachel Carson's *Silent Spring* in 1962 and a number of distressing environmental disasters were beginning to call attention to the impact of industrial capitalism on the natural environment. The Club of Rome's 1972 report *Limits to Growth* brought into question assumptions of never-ending economic growth given the physical limits of natural resources, and the wavering ability of the planet's land, air, and water to absorb increasing amounts of industrial, energy, and consumer waste. It was at that point, not long after the celebration of the first Earth Day, that environmental sociology began to come into its own as a sub-discipline within sociology.

Allan Schnaiberg's *The Environment: From Surplus to Scarcity* (1980) made clear the structural dynamics that undergird growth-based economies. Namely, the guiding principles in a capitalist system are economic, not ecological, meaning, to a large extent, we have fooled ourselves into believing we need not obey some of the basic laws of physics. In fact, not only is our economic system utterly dependent on the natural world to source its material inputs, but the system we rely on actually accelerates this dependence. With each innovation in production and technology, we become all the more efficient at producing goods and services which in turn require greater amounts of material input. Our current environmental crisis is what happens when systems in relative equilibrium – i.e., the world's ecosystems – are severely disrupted by growth-based economies. Schnaiberg's "treadmill of production" theory effectively turned the sociological gaze back at the natural resource origins of capitalist society.

Sociologist William Catton provided a related structural account of human society's dependence on finite natural resources, and developed with his colleague Riley Dunlap, a more value-based foray into why some people seemed to care more about environmental issues than others (Catton, 1982; Catton Jr & Dunlap, 1980). Structurally speaking, it made sense that societal interest in environmental issues ebbed during the mid-20th century. In the United States, post-war economic growth was at a zenith, and households on average experienced remarkable improvements in their material well-being (Gordon, 2017). Also, the population continued its shift away from rural areas, and the distance between people and the natural world, not to mention the toil of agricultural work, grew more vast. This contributed to what Catton and

Dunlap referred to as the "human exemptionalism paradigm" (HEP); that is, a general sense that our intelligence, specialized skills, and mastery of technology made humans to a large degree exempt from the natural resource limitations and constraints set for all other living things on Earth.

There was clearly a perceptive disconnect between the many comforts provided by modern life and the material basis, sourced entirely from nature, that made it all possible. By the 1970s, some sociologists feared this form of blindness would ultimately result in ecological catastrophe. Only a broader consciousness of the environmental crisis which understood our reliance on the Earth's ecosystems for our survival could provide the political will necessary to bring about socioecological change. A "new environmental paradigm" (NEP) was thus proposed as a sociological tool for measuring and understanding this more enlightened predisposition (Catton Jr & Dunlap, 1980).

Why read this book

Despite these and more recent developments, environmental sociology has occupied a space more or less on the outskirts of mainstream sociology. Given its focus on social class, race, gender, institutions, and social networks, you might even argue sociology has been negligent of the world's environmental crisis. In *Sociology Saves the Planet*, I aim to correct this misperception. First, I want to draw out the origins of sociology in the socioecological transformations of the industrial revolution. As we have already seen, Emile Durkheim, Karl Marx, and Max Weber provided early insights into the changing nature of work and modern life as urban denizens became increasingly distanced from their rural, agricultural origins.

The core of this book will explore socioecological perspectives developed since that time which highlight the inextricable connection between environment and society. Special attention is devoted to the dual role of people as producers and consumers in the modern context. Also, reflecting recent developments within environmental sociology, this book examines the significance of major categories of social difference vis-à-vis the current environmental crisis. In that regard the question of environmental justice is paramount, illuminating both the disproportionate benefit of natural resource exploitation to those countries and individuals with higher socioeconomic status, and the greater exposure to environmental hazard among those with less.

Though there are a number of books available for courses on environmental sociology, there is not one I have come across which integrates the theme of environment and society as a core element within an introductory text. This follows a history within mainstream sociology of maintaining ecological issues as a topic outside the primary concerns of the discipline. Many universities, like my own, now require that environmental curricula be taught within introductory courses across academic areas. Moreover, there is an urgent need to underscore the intrinsic value a sociological perspective brings to our understanding of the current ecological crisis – namely, the central themes of sociology are inherently tied to and entangled with our dependence and impact on the natural environment.

Before laying out the plan for the book, I would like to briefly return to the topic of rip currents: another curious characteristic of this phenomenon is that, if you are patient and willing to tread water for a bit, the current's circular flow will eventually take you back to shore with little struggle. I am not saying that what follows in these pages will be effortless to absorb and process. I do want to demonstrate, however, that some of the essential tools we need for understanding the nexus of human society and the natural world have been available to sociologists for a while; the time is ripe to use them for bringing the environment back to the core of sociological analysis. To that end, I would like to highlight eight sociological currents that can draw us back to the human origins of the present ecological crisis, as well as direct us to a more balanced, sustainable future. Not coincidentally, these are also central themes in the chapters that follow.

Central themes and plan of the book

First, *rationalization* is the term Max Weber used to describe something especially distinctive about life under capitalism. To be clear, this is not what happens when you justify eating dark chocolate because you read somewhere that antioxidant flavonoids are good for you. No, Weber's rationalization refers to the tendency among capitalists to seek out increasingly efficient and predictable means to produce goods and services, all in the name of higher profits. This impulse towards profit and capital accumulation, according to Weber, had specific cultural origins in 17th-century Protestantism (Weber, 2013). Though its religious derivations have long since faded, Weber believed this now dominant cultural disposition within capitalism was leading modern societies down a bleak and dreary corridor toward what he referred to as the "iron cage." In contrast to preindustrial times when the world remained "a great enchanted garden," cold calculation, control, and profit are now the driving factors in modern life (Weber, 1993).

Little wonder the world's ecology has been sidelined: the dominant cultural frame places people and nature within a matrix that prioritizes above all else lower costs and greater financial return. In Chapter 2, we will look at specific examples in transportation, agriculture, and energy where the culture of rationalization has resulted in some profitable yet socioecologically undesirable, at times disastrous, outcomes. Once we can identify the common theme of rationalization and recognize it as a cultural trait particular to our moment in history – not inherent to human nature – alternative systems of value that prioritize sustainability and human dignity seem more like a real possibility.

The next sociological theme covered in Chapter 3, *socialization*, underscores the possibility that society and individuals within it can and often do change, given the right social context. You may have at some point found yourself in a maddening argument with someone about the relative import of nature versus nurture; i.e., was our lot in life genetically predetermined, or did life experience, trauma, and chance ultimately determine our fate? The answer, of course, is both; we inherit both physical and personality traits from our parents and ancestors, and we clearly also acquire knowledge and patterns of behavior from lived experience. Sociology – along with its sister discipline anthropology – places considerable weight on understanding the learned behavior side of social life. We would not be who we are if not for the others that surround us – the complexities of spoken language being the most obvious example in this regard.

There are other categories of humanhood, however, which, though on the face of it may seem genetic and fixed, are also themselves inhabited and learned in ways that require exposure to socially determined culture. Sex and race, for example, may seem like categories you are born into. It does not take much scratching below the surface, however, to see that they also originate in culture and can take on meanings and labels that, in a hierarchical society, give advantages to some and disadvantages to others. By giving a name to this process, however, we are also providing space to begin changing patterns of racism, sexism, dominionism over nature, and other prejudices we have all to one degree or another been socialized into.

Socialized learning can happen throughout the life course, though it is clear that childhood is a particularly important time to instill through education and by way of example socioecological values. One of the great challenges in this regard, as we will see, is that marketers and advertisers have long been privy to the power of socialization and, when we neglect the values we wish to instill in our children, they are more than happy to step in and socialize them on their own terms, typically with an eye for extracting profit.

In Chapter 4, we will examine the concept of *multiplexity* as a way to bridge the gap between environment and society. It comes from the literature on social networks, but its theoretical origins go further back, signaled originally in Emile Durkheim's conceptualization of *social integration*, by which he meant people's complex interdependence with other people. Multiplexity operationalizes social integration by describing a network where people know each other in multiple ways (Verbrugge, 1979). So, not only are you my friend, but our kids attend the same school, we are in the same book club, and our dogs regularly meet up at the same dog park. Some of this may seem trivial, but knowing the same people through different activities and different organizations is a precious source of trust and solidarity in modern life.

Do you ever feel like it is not worth it to even try to behave in a sustainable, socially conscious way if no one else is doing the same? Unfortunately, you might be right. As we will see throughout this book, very little happens in the way of social change without coordinated, collective behavior among a critical mass of people. A major corollary to this is the truism that people are not likely to work together for a common cause unless they trust the other people with whom they coordinate. Regarding socioecological change, just remember the next time you tutor disadvantaged kids, volunteer at the food shelf, or join a pick-up basketball game, you are also working to create multiplex trust in a society that would do well by connecting more across different categories of people.

Chapter 5 takes up the classic sociological term of *alienation*, drawing our attention to the gaping distance between us and the things we consume; specifically, how, by whom, and under what circumstances they were produced. Most of us are truly out to sea when it comes to understanding how much people were paid to assemble our smartphones, where electricity comes from when we flip a switch inside our homes, and how we would survive without government subsidized industrial agriculture. Alienation also refers, however, to our relationship to each other. Since when, for example, did the role of consumer start to displace our role as citizens? The former defines a person with a set of individual needs who depends on the market to fulfill them through the exchange of cash payment. The latter describes people with personal needs, for sure, but also with a degree of responsibility to each other and for the direction society itself will take. Almost by definition, a consumer society requires more participation in the acquisition of goods through market exchange, while leaving civic participation by the wayside.

The problem of course is the less connected and committed people are to the common cause of managing a society, the more society-wide decisions will be determined by the market and the underlying drive for profit. Socioecological change thus requires we shift our focus away from the universe of consumer products towards our meaningful relationship with other people, a point underscored by the fact that our connection to nature itself is often mediated through our relationship with other people.

In Chapter 6 we will examine the social network concept of *homophily* as a way of introducing us to the core sociological theme of *social inequality*. Homophily captures the tendency in modern life to gravitate to people like ourselves, a seemingly mundane fact which presents some challenges to both social science research and society, in general. For researchers, homophily reminds us that we need to be particularly careful about the assertions we make based on the data available to us. Cause and effect in social life can be hard to disentangle when people are already to some degree presorted on the basis of common interests, education, culture, and socioeconomic status. This does not undermine the sociological project altogether, but it does prompt us to be careful and clear about the limits of social science data we have at our disposal.

At the experiential level, life in a homophilic society can lead to distorted perceptions of the way the social world actually operates. When we are surrounded by people similar to ourselves, we may start to feel that our way of life is normal, neglecting the fact that there is a universe of difference, ways of being, and beliefs different from our own. Homophily eases our social interactions by surrounding us with similar others, while simultaneously buffering us from the diversity and adversity of people different from us. The sociological study of inequality is thus an essential starting point for us to be able to step outside ourselves and empathize with others whom we might not encounter on regular basis. This can be done both quantitatively through survey research on income, education, occupations, and wealth, and through more qualitative methods that capture lived experience and the meaning of social life through face-to-face interviews, ethnographic observations, and direct participation in organizations, households, and places of work.

Social construction is a concept sociologists use to describe how categories of difference such as race, class, and gender come into being. During the social upheaval and tumult following the murder of George Floyd in 2020, for example, my eight-year-old son asked me in all earnestness why the Black Lives Matter movement was not called All Lives Matter. "Don't we all matter?" This was a great question as well as a good introduction to the notion of false equivalencies. Black lives matter in particular to this movement because, beginning in 1619 when the first African people were enslaved by European colonists and brought to the Americas, Black lives have been consistently abused, exploited, and neglected, in a society dominated by whites. Black lives matter because for too long they have not mattered in an unequal and racist society. As this example makes clear, to say something is socially constructed does not make its effect on people's day-to-day lives any less concrete – quite the opposite. Social construction calls attention to how deeply rooted in history and human behavior the biases are that perpetuate injustice.

For some, addressing social justice and inequality may seem a lower priority or even antithetical to addressing the global environmental crisis. What good is a more just society if there are no suitable places left on the planet to live? By taking social inequality seriously, however, sociology invites us to rethink the equation of sustainability versus a more equal society. Research on *environmental justice* – the sociological theme taken up in Chapter 7 – shows us that some parts of the population bear disproportionately the weight of environmental hazard. This is evident at the regional level where African Americans, American Indians, Latinxers, and the poor are on average much more likely to live proximate to waste incinerator plants, refineries, and other polluting industries than is the majority population (Bullard et al., 2007). Cross-nationally, the impact of climate change has up to now been concentrated in the global south, where desertification and the loss of arable farmland is most acutely felt (Rigaud et al., 2018). When looked at from the point of view of who pays the social and environmental costs of industrial capitalism, it becomes clear that a more sustainable future will not be possible without first addressing the massive inequities that both provide consumer comforts for some and environmental calamity for others.

The *perception of risk* and how that influences whom and what we trust is the next sociological theme taken up in Chapter 8. This is of critical importance in the contemporary political environment where topics seem to toggle back and forth between hope and fear. The scapegoating of vulnerable groups for society's woes is always a temptation for politicians who wish to draw attention away from more pressing economic or social issues. Non-citizen immigrants, in part because of cultural differences but also because of their inability to vote, are easy targets in this regard. Everything from terrorism, to violent crime, to even increased environmental risks have been blamed on resident immigrant populations. This is remarkable because sociological research demonstrates repeatedly that immigrants are relatively less likely to either commit acts of terrorism or engage in violent criminal behavior, and their carbon footprint is small when compared to that of native-born Americans (Head, Klocker, & Aguirre-Bielschowsky, 2019; Light & Miller, 2018; Nowrasteh, 2016).

In fact, the question of how our attitudes and beliefs relate to reality has long been a challenge within environmental sociology. A consistent finding in this regard is that people's expressed environmental attitudes often have a tenuous association with their actual behavior. For example, the vast majority of Americans are aware of climate change and would like to do something to slow down its progress (Funk & Kennedy, 2020). Yet, year after year, the amount of carbon dioxide and other greenhouses gases emitted into the atmosphere increases. Though discouraging, this line of research makes clear that changing people's attitudes towards socioecological problems is probably much less important than regulating industry and changing the structure of unthinking consumer practices that result in what most of us agree are undesirable outcomes.

Socioecological change is the final theme we take up in Chapters 9 and 10, directly tied to the question of social structure. Structures, as the word implies, are not easy to change; they are solid and support the many layers of culture, social networks, and economic organization that constitute our society; and yet, a more equitable and sustainable society requires that they change. There are, of course, superficial, or symbolic changes that may provide the appearance of change. What is needed, however, to draw us away from the directional flow towards imminent catastrophe is the redistribution of opportunity and the reorganization of production in ways that promote human dignity while respecting our interdependence with the natural world. We will look at the literature on social movements which shows how protests and non-institutional politics have worked effectively in the past to draw attention to long-standing grievances held by minority groups in the population and which have actually resulted in significant structural changes.

We will also consider the increasingly important role the forces of nature play in human society. No serious plans for the future can be laid out without taking into account the impact natural disasters and zoonotic threats now have on modern life. Political activism during the Covid-19 pandemic reminds us that a shared sense of solidarity is still essential for a social movement to gain traction in a diversified media context vying for people's attention. Before solidarity, however, people must trust each other, and before people trust each, they must also have the opportunity to engage with a variety of people in a diverse society. Homophily – i.e., the tendency for people to know people similar to themselves – is a major obstacle in the way of creating general trust within a diverse society. Fortunately, sociology has some suggestions for how to turn that around.

In the concluding chapter we will consider five distinctive socioecological strategies that can be used within the broader effort to bring about lasting, positive socioecological change. In concert with technological innovations and research in the natural sciences, sociology reminds us it has been the co-ordinated actions of human beings which have sucked us out towards a mael-strom of ecological trouble, and it is only the deliberate, thoughtful actions of cohesive human groups that can get us safely back to shore.

Glossary

Alienation   An idea developed by Karl Marx, alienation is the lack of meaningful relationships of workers with other workers and with the things they produce in a capitalist society.

Communist Revolution   The people's movement to eventually put an end to class inequality characteristic of capitalist society; a key element of Marx's theory on capitalism.

Environmental Justice   The social movement which calls attention and seeks equitable solutions to the concentration of environmental hazards, including exposure to air and water pollution, and proximity to toxic industries, among communities of color and the poor.

Homophily   The tendency of individuals to know and build relationships with others who are similar to themselves. This pattern both polarizes and unites groups of people in society.

Multiplexity   A social network context in which individuals know the same people in different ways – e.g., neighbor, coworker, fellow gardener. Multiplex networks are associated with highly levels of trust among their members.

New Environmental Paradigm   William Catton and Riley Dunlap's theory describing a disposition which recognizes both society's depend-ence on nature for its existence, and the role the natural environmental plays in determining human activities.

Organic Solidarity   An element of Durkheim's social theory which cap-tures the modern, complex, and interdependent social context wherein people with specialized skills rely on others with their own specialized set of skills for their mutual survival.

Rationalization   The tendency within capitalism to seek out increasingly calculable, effective, and predictable means to produce goods and ser-vices for profit.

Scapegoating   The deliberate blaming of a class of people for society's problems, often based on legacies of cultural prejudice held towards groups disparaged by those in power.

Socialization   The acquisition of culture, norms, and behavior through regular social interaction with others; a process particularly evident dur-ing early childhood, but which occurs throughout the life course.

Social Construction   An idea, concept or category originating in human history and culture with concrete social implications in our everyday lives; salient examples of this include race, gender, and socioeconomic status.

Social Darwinism   Herbert Spencer's application of evolutionary theory to the social world, upholding the idea that those who have the most social strength and resources at their disposal are better equipped to be members of society than those without.

Social Inequality   Advantages or disadvantages granted to certain groups in society that are inherent to greater social patterns and not random, arbitrary choices made by individuals.

Social Movement   A collective effort to voice a shared grievance or perceived injustice and bring about social change, usually through actions taken outside institutional forms of political process such as voting and court rulings.

Treadmill of Production   An idea based on the research of sociologist Allan Schnaiberg, describing how the more effective capitalism is in deploying efficient technologies to produce its goods and services, the greater the strain is placed on the natural environmental.

Verstehen   A term to signify a deep understanding when examining social research. *Verstehen* is an idea developed by Max Weber encompassing the awareness that studying humans as social subjects may imply biases as the researchers are humans too.

Zoonotic Threats   Viruses, bacteria, parasites, and other miniscule biological entities which have and continue to play an important role in determining the fate of human societies. They are often passed on from other animal species to human beings.

Questions

1   Was there a time in your life when a regular experience in nature structured your life? Was it in a city? Your backyard? During summer vacation, perhaps? How did this experience change the way you think about nature? After that experience, did you end up valuing all the more the luxuries of modern life, or did this spark a desire in you to spend more time away from the comforts of home? Why?

2   From its origins in the 19th century, what was sociology's relationship with the natural sciences? What influence did Charles Darwin's theory of evolution have on social theory? In what key ways did the works of Karl Marx, Emile Durkheim, and Max Weber differ from Herbert Spencer's ideas on the "survival of the fittest?" Do you think "social Darwinism" continues to influence people's perception of society today? Give an example or two.

3   What might explain the "decoupling" of American sociology from anything having to do with the natural world during the first half of the 20th century? Why do you think environmental sociology began to emerge as a unique area of study when it did during the 1970s? Despite increasing interest, why do you think environmental sociology continues to be on the margins, not in the center of mainstream sociology along with topics like race, gender, social class, institutions, and social networks?

References

Bullard, R. D., Mohai, P., Saha, R., & Wright, B. (2007). *Toxic wastes and race at twenty 1987–2007: Grassroots struggles to dismantle environmental racism in the United States*: United Church of Christ Justice and Witness Ministries.

Carson, R. (1962). *Silent spring*: Houghton Mifflin Harcourt.

Catton, W. R. (1982). *Overshoot: The ecological basis of revolutionary change*: University of Illinois Press.

Catton Jr, W. R., & Dunlap, R. E. (1980). A new ecological paradigm for post-exuberant sociology. *American Behavioral Scientist, 24*(1), 15–47.

Darwin, C. (2003). *On the origin of species, 1859*: Routledge.

Durkheim, E. (2014). *The division of labor in society*: Simon and Schuster.

Engels, F., & Marx, K. (2004). *The communist manifesto*: Broadview Press.

Funk, C., & Kennedy, B. (2020). How Americans see climate change and the environment in 7 charts. *Fact Tank: News in the Numbers*. Retrieved from https://www.pewresearch.org/fact-tank/2020/04/21/how-americans-see-climate-change-and-the-environment-in-7-charts/

Gordon, R. J. (2017). *The rise and fall of American growth: The US standard of living since the civil war* (Vol. 70): Princeton University Press.

Head, L., Klocker, N., & Aguirre-Bielschowsky, I. (2019). Environmental values, knowledge and behaviour: Contributions of an emergent literature on the role of ethnicity and migration. *Progress in Human Geography, 43*(3), 397–415.

Light, M. T., & Miller, T. (2018). Does undocumented immigration increase violent crime? *Criminology, 56*(2), 370–401.

Meadows, D. H., Meadows, D. L., Randers, J., & Behrens, W. W. (1972). The limits to growth. *New York, 102*, 27.

Nowrasteh, A. (2016). Terrorism and immigration: A risk analysis. *Cato Institute Policy Analysis*, 798, September 13. Retrieved from https://www.cato.org/policy-analysis/terrorism-immigration-risk-analysis. Accessed September 16, 2021.

Parsons, T. (1949). *The structure of social action* (Vol. 491): Free Press.

Rigaud, K. K., de Sherbinin, A., Jones, B., Bergmann, J., Clement, V., Ober, K., Schewe, J., Adamo, S., McCusker, B., Heuser, S., & Midgley, A. (2018). World Bank.

Schnaiberg, A. (1980). *The environment: From surplus to scarcity*: Oxford University Press.

Spencer, H. (1969). *The man versus the state, with four essays on politics and society*: Penguin Group: New York.

Verbrugge, L. M. (1979). Multiplexity in adult friendships. *Social Forces, 57*(4), 1286–1309.

Weber, M. (1978). *Economy and society: An outline of interpretive sociology* (Vol. 1): University of California Press.

Weber, M. (1993). *The sociology of religion*: Beacon Press.

Weber, M. (2013). *The Protestant ethic and the spirit of capitalism*: Routledge.

2 Efficient, rational plans and unintended socioecological outcomes

A naïve account of the relationship between efficiency and the environment is perhaps best represented by the Shel Silverstein classic children's book, *The Giving Tree* (1964). In it, an apple tree befriends a small boy whom she happily lets climb her branches and eat her apples. Early on, the tree and boy spend a lot of time together creating a close cross-species bond. As the years pass by, though, the boy grows older and begins to visit the tree only when he needs something from her. First, it's apples to sell for money, then branches to build a house, and then the tree's trunk to build a boat. With each additional demand placed on the tree, the "boy" (now a man) gets what he wants and the tree both abides and is happy, until in the end there is nothing left but a stump for the aged protagonist to sit on.

Analogously, our experience as human beings in industrial capitalism is much like that of the boy. Whereas, early on we were just eating the apples and inhabiting natural space, our demands on the environment began to increase as we became more efficient at extracting resources from the natural world, eventually surpassing, as many have argued, the Earth's natural limits. Fortunately – and herein lies the naïve part – the Earth is always there when we need her and is happy to supply us with the material basis of our own happiness willingly and without complaint. Efficiency from the *Giving Tree* perspective appears to make perfect sense to the man/boy who, through the course of a lifetime, exhausted all the natural resources at his disposal. No one, it seems, thought to ask the tree.

Beginning in the late 19th century, German sociologist Max Weber put a lot of thought into efficiency and other aspects of capitalism that made modern life distinct from what had come before. A key concept and characteristic of this new period of human history was what he called *rationalization*. His use of this term, it should be pointed out, is quite different from its popular psychological usage – a way of justifying after the fact an action we have already taken or that has already occurred. Instead, what Weber meant by rationalization was the tendency in modern life to base decisions on rational calculation, efficiency, and increased predictability in desired outcomes (Weber, 2013). As he saw it, these rational motivations were displacing traditional values and customs as dominant forms of understanding and the bases for taking action in the world.

DOI: 10.4324/9781003110668-2

Vehicle emissions and the ethical limits of Weber's rationalization

A central irony in the dominant culture of rationalization is that when we give cold calculation and cost reduction precedence over broader concerns affecting society and the ecology we leave the door open for irrational and even unethical outcomes. To give a personal example, after the birth of our second child in 2011, my wife and I felt it was time to invest in a new a car. We made up our minds to own only one car, so long as it met our strict standards. Namely, it had to get good gas mileage (efficient), have a good safety record (predictable), and – so we could still maintain our sporty attitude – it had to have a stick shift (okay, maybe not entirely rational). We homed in on the Volkswagen Jetta SportWagen.

At the dealership, though we already had our eye on the base SE model, which ran on standard grade gasoline, the salesperson suggested we try the SportWagen TDI which, though identical in appearance to the SE, ran on diesel fuel. This meant not only would we pay less on gas because of its better mileage, but it would also have better pick up and stronger torque, thus adding to its sporty appeal on the road. Our test drive confirmed this last claim. I seemed, however, to remember hearing long ago that diesel engines were far more polluting than those in regular cars. This had stuck with me because I also recall learning that the reason exhaust from large commercial trucks was more polluting than that from other vehicles was because an exception in federal law allowed them to run on diesel fuel. When I mentioned this to the salesperson, he assured me the technology had advanced and diesel engines were actually as clean or cleaner than those in other non-commercial vehicles. My wife and I were nearly persuaded – we could have fuel efficiency, high torque, AND pollute less with a diesel-powered car. We did, however, want to stick to our budget and the TDI was a few thousand dollars out of range. We bought the SE model and never looked back.

That is, until a few years later in fall 2015 when we and the rest of world discovered that the Volkswagen Corporation had since 2009 been deceiving the public and regulators around the world, leading them to believe their diesel vehicles met environmental standards for air pollution. One television commercial at the time included VW engineers sprouting angel wings as they inspected newly minted, low polluting diesel vehicles. In contrast to their saintly depiction in this high-priced Super Bowl ad, what VW engineers had actually accomplished is devise onboard computer technology which, when sensing a vehicle was being tested for emissions, set the diesel engine into a less torquey "safe mode" that spewed out much lower levels of greenhouse gases, specifically, nitrogen oxide. When running in its regular mode outside the emissions test setting, the VW TDIs emitted as much as 40 times the amount of permitted levels of nitrogen oxide into the atmosphere (Ewing, 2017).

Worldwide, some 11 million TDI vehicles were put on the market over a six-year period. Clear discrepancies between emissions tests done in the lab versus those done on the road were brought to global attention by researchers at the University of West Virginia in September 2015. VW was forced to acknowledge its deceptive scheme, VW stocks plummeted, and VW CEO Martin Winterkorn resigned. Perhaps even more disturbingly, as many as six other major carmakers came under investigation for similar ploys involving their own diesel vehicles (Carrington, 2015).

Why would a global enterprise such as Volkswagen put its reputation at stake in this way? One important factor was the policy environment which increasingly took into account the growing threat of climate change. The pressure was on from national governments – including that of the United States – for industry to rise to the occasion and meet this threat with energy efficient, low polluting technologies. One tack taken by car companies was to begin the transition to electric vehicles, including engineering greater storage capacities for automobile batteries. Another approach was to work within the limits of the internal combustion engine, maintaining the still unchallenged power advantages of this technology while reducing pollution. In the late 20th century, there had been some early indications that the diesel engine might be modified in a way that would pollute less. By 2009, however, it was clear it would not be up to the task of meeting rigorous emissions standards. Expectations were high among both regulators and stockholders and, though the details are not yet clear, it is evident that at some point VW executives calculated they could skate by on deceptive emissions test results until a truly less polluting technology actually came around (Ewing, 2017).

Weberian rationalization predicts VW might have been tempted to neglect other ethical, social, and ecological considerations in order to attain its over-riding goal of profit maximization. By not announcing to the world that they were unable to meet the exigencies of environmental policy, VW ostensibly met the pressing standards of predictability, calculability, and efficiency that investors and consumers could both buy into, even if it did so on the bases of fallacious test results.

The unintended outcomes of smartphone technologies

The follies of rationalization in some ways define the modern age. The drive for predictability, calculability, and efficiency may result in what Weber re-ferred to as the "paradox" of unintended consequences; that is, the unantic-ipated outcomes resulting from our (usually well-intended) actions (Weber, 1978). To some degree this is a matter of cognitive load – the very act of focusing our attention on a specific goal causes us to screen out other possible or concurrent outcomes.

The smartphone is emblematic of this dynamic. The celebrated merging of communication and micro-processing technology in this device, though facilitating an array of efficiencies in our personal lives, may on occasion con-flict with other things we value as a society such as personal safety, privacy, and a healthy, meaningful childhood.

Smartphones have undeniably streamlined our lives. Through the hi-tech integration of microchip computing and telecommunications, smartphones provide us access to the internet practically anywhere. Phone conversations, emails, texting, social planning, music, videos, high-resolution photography, GPS, online shopping, current events, and reviews of everything from movies to tourist destinations to college professors are readily available whenever, wherever we need them. Never before in the history of the planet has so much information been available to so many people. The possibilities, it would seem, are endless though not without their own set of unintended consequences.

One troubling aspect of smartphone technology is mounting evidence that it is addictive and can lead to compulsive and even lethally dangerous behavior. Regarding the latter, "distracted driving" is a growing problem, the cause of approximately 10 percent of vehicular fatalities in the United States, today. According to the National Highway Traffic Safety Administration (National Highway Traffic Safety Administration, 2019), the observed "manipulation of handheld devices" by drivers has been trending from 0.5 percent in 2006 to 2 percent in 2017 in the general population. This observed tendency has been much more pronounced among 16- to 24-year-olds for whom this behavior has increased from 0.4 percent to 4 percent during the same period after peaking out at 4.9 percent in 2015. To be clear; it is not that 4 percent of 16- to 24-year-olds sometimes use electronic devices while they drive but rather, at any given moment on average, 1 out of 25 young drivers on the road are focused on a technology that is distracting them from driving.

Socially, there is much ongoing debate about the impact mobile devices have on our connections to others. One of the main concerns is that our increased dependence on smartphones and social media as our primary interface for social interaction is that we will no longer know how to interact with people on an interpersonal basis. For most of us, this particular concern seems to be overblown – much research in this areas suggests our online and cellphone based networks reflect directly our face-to-face relationships; i.e., the people we are most likely to interact with online are also the people we are most likely to meet with in person (Boase et al., 2006). There is, however, evidence that some people are being left behind. If you are a gregarious extrovert, for example, who enjoys the company of people, it is likely that your online/mobile connections will only enhance your sociability. On the other hand, if your personality is more inward and you find it hard to interact casually with others, life online may provide you a protective cocoon that is difficult to poke through. There is a good argument to be made that electronic media tend to amplify our pre-existing dispositions towards the social world.

Another angle on this is to draw attention not so much to what we do when we are staring at a screen, but to all the things we are not doing because we spend so much time using digital technology. As adults, it is easy to assert the defense of digital freedom and, as just mentioned, what we do online and on our phones mainly enhances, for better or for worse, what we would already be doing, anyway. When it comes to children, however, the question of childhood socialization and its impact on human brain development brings in a whole different set of concerns regarding unintended consequences.

Across the age spectrum, we know that kids are spending less time playing outside. We will consider this matter in more detail in the chapter on socialization. Suffice to say, digital technology has dramatically transformed the experience of childhood over the last two decades. Among some groups in society – including most notably the families of CEOs and engineers who have worked hardest at integrating online and mobile communication into our daily lives – deliberate efforts have been made to keep their children away from these devices for long as possible (Bowles, 2018).

The historical trend towards rationalized agriculture

Industrial agriculture is another area where we have literally reaped the benefits of rationalization through the application of scientific research, increasing crop yields to levels capable of feeding some 7-plus billion people on the planet. To gain some perspective, it is worth remembering that for most of our 200,000-year history as a species, agriculture did not exist at all. Until about 10,000 years ago, homo sapiens eked out a living solely as hunter-gatherers (Panter-Brick, Layton, & Rowley-Conwy, 2001). They foraged, eating and preparing food in the wild where they could find it, and if they could not, they picked up and moved somewhere else. We were (and, to some degree, still are) a restless wandering species who, by the end of the pre-agricultural period, made our way around the world, including the far-flung continents of Australia and the Americas.

Then, around 8000 BCE, a couple of millennia after the most recent ice age, something happened in the region we now think of as Iraq and Syria, also referred to as the Fertile Crescent. True to its name, this large geographic swathe centered on the space between the Tigris and Euphrates Rivers was ideal for settling down and farming (Von Soden, 1994). Relatively large permanent human settlements emerged in this region for the first time. Humans continued to migrate all over the globe, but now with the advent of agriculture, they did not have to. In fact, as farming became increasingly refined and efficient, a small percentage of people did not have to labor to produce food at all. That is, when done efficiently, agriculture created a surplus of food for the community. This meant that, for some, the time and energy that might have once been devoted to foraging for nuts and berries, and hunting wild beasts, could now be applied to other ends.

As surpluses increased and populations grew, the management and distribution of agricultural resources became all the more important, creating new challenges. On what basis should people be allotted food surplus? Should people with greater responsibility in an increasingly complex society be awarded greater status? How will the symbols of that status be best conveyed? And, most practically, how are we going to keep track of all this? Not coincidentally, the earliest evidence of the written word also comes from this region of the world. Clay inscriptions of the Sumerian language date back to about 3000 BCE (Schmandt-Besserat, 2010). As much as a thousand years prior, however, numerical writing had been used as a basis for record keeping, clearly tied to the exchange of goods and the management of agricultural production. Food surplus and writing are thus inextricably linked – the inventory of comestible/tradable goods required a written system of accounting.

For millennia, then, the work of most humans on Earth was directed towards the production of food, mainly through agriculture. As recently as the end of the 19th century, most Americans held jobs somehow related to agriculture. As recently as the end of the 20th century, the majority of people on the planet lived in non-urban settings (Massey, 2005). One of the dominant demographic trends over the last few hundred years has been the movement of people away from agricultural labor and rural life into towns and cities. The acceleration of this trend over the past century can be attributed in large part to the exploitation of a finite resource we have become increasingly efficient at extracting from deep below the Earth's surface – fossil fuels.

Fossil Fuels and our opulent society

In a hundred-year period, fossil fuels have utterly transformed modern life, injecting the rationalized culture of profit maximization with an unparalleled boost, notably as the principal source of fuel in transportation and industry; in the synthesis of pesticides and fertilizer for industrial agriculture; and as feedstock for plastics in the production of practically everything. Our societal dependence on petroleum, coal, and natural gas since the mid-20th century has been near absolute. And for good reason: The relative luxuries of modern life are almost entirely reliant on our consumption of fossil fuels. The most direct way to understand this is through our relationship with energy. Consider, for example:

- The gasoline that fuels both personal vehicles and the commercial fleets that distribute the goods we depend on for sustenance and distraction
- The methane gas that cooks our food and heats our homes
- The coal used abroad in the production of consumer goods, and which still generates a fifth of all electricity consumed in the United States for lighting, refrigerators, microwave ovens, air conditioning, and the charging of phone, computer, and electric car batteries, among other things
- The kerosene that fuels the airplanes which bring the world to us and us to the world in ways that were unimaginable a mere century ago.

Systems theorist Buckminster Fuller suggested decades ago that we might better understand our dependence on the highly concentrated form of energy we receive from fossil fuels if we put it all in human terms. The term he used for human equivalents in energy output is "energy slaves" (Fuller, 2001). For example, to power a 100-watt light bulb for an hour would require the continuous effort of five human beings or energy slaves. This way of thinking, Fuller argued, might make us think twice about leaving an extra light on in an empty room which, for an hour, consumes the equivalent of a pound of coal. Other relevant comparisons include speeding down the interstate at 75 mph (2,000 energy slaves) or using an elevator in a ten-story building (100 energy slaves).

The point of all this is we have grown accustomed to technologies that consume enormous amounts of energy without us even thinking about it. This is largely because fossil fuel consumption is built into the physical and social structures of our lives; stepping away from them will require much more than a change in attitudes or increased levels of concern about the natural environment. Fossil fuels have thus provided the ultimate rationalized enhancement to productivity and efficiency at an incredibly reasonable cost – much lower, as the energy slave analogy reminds us, than paying for the equivalent output through human effort. Their impact on the environment, however, is another story altogether.

Unintended consequences of fossil fuel consumption

Back in the 1950s and 1960s when the US economy saw its highest levels of growth, the United States was a net exporter of petroleum, drilling readily accessible oil from wells in the fields of Texas and California, and to a lesser extent off the shallow shores of the Gulf of Mexico. In 1970, however, the United States hit a peak in its production of domestic oil. This did not mean that we were going to run out of petroleum any time soon, but rather it was becoming more expensive to extract from the earth that which remained. Increasingly, we came to rely on far away nations for our domestic oil consumption. When our support of Israel during the Yom Kippur War of 1967 led to an oil embargo against the United States from countries in the Middle East, it became clear that new strategies were required to get the energy we needed closer to home. Unfortunately, most of these strategies also incurred higher costs for extraction, both financially and environmentally.

The most successful find in the United States since that period was in northern Alaska's Prudhoe Bay, the largest oil field in North America. This was a godsend for the energy economy but it also meant working in some of the most challenging offshore conditions imaginable north of the Arctic Circle. Moreover, petroleum drilled from this far northern region of Alaska would somehow have to make it to the energy-hungry consumer markets of the lower 48. The solution – the trans-Alaska pipeline, constructed between 1974 and 1977, stretching from Prudhoe Bay in the north to the port town of Valdez 800 miles to the south. Construction of the pipeline met with much resistance, significantly from indigenous groups who saw this as an outside incursion on native lands. At the same time, environmentalists pointed to the inherent risks of oil spills and contamination on animal habitats and the unspoiled natural landscape, including streams and rivers, and the fragile permafrost surface of the Alaskan tundra. An additional concern was the next leg of transit from the southern end of the pipeline to the seafaring oil tankers that brought the crude unprocessed product down south. It was precisely at that point in the journey that the highly rationalized extraction and distribution of petroleum from Prudhoe Bay resulted in the unintended consequence of environmental calamity.

A little after midnight on March 24, 1989, the Exxon Valdez, an oil tanker carrying crude oil drawn from the trans-Alaska pipeline, ran into a reef in Prince William Sound, just off the coast of Valdez, Alaska. Pouring approximately 11 million gallons of petroleum into the ocean and spreading over a thousand miles of pristine coastline, the Exxon Valdez oil spill was at the time the largest human-caused environmental disaster in American history (Tierney & Quarantelli, 1992). Laser focus on the siphoning of energy from the earth to supply power and raw materials for profit led to a disastrous unintended consequence. Keeping with the theme of rationalization, it is worth noting that, in an attempt to cut costs, the number of people working on the tanker's crew had been cut in half over the previous ten years. Additionally, though reef and iceberg sensing technologies were available to prevent exactly this type of event from occurring, they were not in use on the Exxon Valdez the early morning of March 24. In that way, a few cost-cutting measures contributed to the deaths of hundreds of bald eagles and killer whales, thousands of sea otters, hundreds of thousands of seabirds, and millions of other animals on land and in the sea whose habitats were devastated by the spill.

You might think and hope that at the very least industry and environmental regulators would have learned some important lessons from the Valdez experience to assure that a preventable disaster like this would never happen again. Unfortunately, 21 years later the Deepwater Horizon disaster in the Gulf of Mexico proved that industry was willing to take even greater risks in the name of greater profits, the environment and nearby human populations be damned. Two days after a methane gas explosion on the Deepwater Horizon oilrig killed 11 people on April 20, 2010, it became evident that, in addition to the loss of human life, a major environmental disaster was underway. Nearly a mile below the surface and 41 miles off the coast of Louisiana, a wellhead had been severely damaged and was spewing out crude oil at a rate of 260,000 gallons of oil a day (Robertson & Krauss, 2010).

It was not until 87 days later that British Petroleum, the well's owner, was able to announce they had finally contained this gushing underwater breach. When all was said and done, approximately 210 million gallons of crude oil were released into the Gulf of Mexico, the largest oil spill in history whose deleterious impact on the Gulf ecosystem and people whose livelihoods depend on it will likely persist for decades. Once again, more could have been done to prevent this from happening. A federal panel charged with investigating the Deepwater Horizon disaster found that there was a "rush to completion" of the rig's construction, and that "there was not a culture of safety" on the drilling rig (Kim, 2017). In their book, *Blowout in the Gulf: The BP Oil Spill Disaster and the Future of Energy in America* (2010), sociologists William Freudenburg and Robert Gramling define the "atrophy of vigilance" as "the expectation of organizational performance to get sloppier over time, particularly in the case of rare or 'unexpected' problems" (p. 35). Each year that goes by without a major disaster occurring is interpreted by industry leaders as confirmation that it is okay to let their guard down and invest less in safety and environmental protection.

The rationalization of personal health

Another area in which rationalization has shaped lives in ways that we are not always individually aware is our health. Approximately a third of all Americans are clinically obese and another third are overweight (Flegal et al., 2016). This is increasingly a public health issue throughout the world: the United States and Mexico are the two nations with the highest rates of obesity in the world. Purely genetic arguments for understanding differences in average weight among the human population do not seem to apply. With regard to places of origin, the United States is one of the most diverse populations on Earth, and yet relatively high BMIs are disproportionately concentrated here. Unfortunately, discounting genetic explanations often shifts the blame for this situation over to individuals who, it is argued, simply need to take greater responsibility for their own health, including eating better and getting more exercise. Both these approaches – genetic explanations and personal responsibility – tend to reduce the obesity crisis to the realm of personal troubles, missing the broader social origins of poor diets and sedentary lifestyles.

Chronologically, we see that obesity as a far-reaching health concern did not really emerge until the late-1980s. According to the federal Centers for Disease Control and Prevention, the average adult American male weighs nearly 30 pounds more now than in the 1960s. What are the factors that have led to this remarkable expansion in the American waistline? Much of it comes from changes in the way we work. Back in 1900, nearly 40 percent of Americans lived on a farm working in agriculture, a sector requiring long hours of manual labor (Lusk, 2016). Today, as the utilization of fossil fuel inputs has greatly reduced the need for human farm labor, barely 1 percent of the workforce is employed in agriculture, a sector of the economy not seen by most Americans as a desirable field in which to work (USDA, 2019). Not coincidentally, about three-quarters of farm laborers are immigrants, while about half are undocumented workers (USDA, 2020). Because of this massive shift away from agriculture, most of us are not engaged in the most grueling forms of manual labor. In fact, the vast majority of Americans work inside, unexposed to the elements, with a good many of us (as I am right now) sitting at a desk in front of a computer screen.

Related to where we work and what we do is how we get there. With the exception of the small minority of us privileged enough to ride our bike to work or to live in city with reliable and convenient public transportation, the vast majority of us commute in cars. For some of us it is a short commute, for others, two hours a day in a motor vehicle shuttling between home and work is not unusual. More time driving means less time doing other more physically demanding activities. More time driving (not to mention more time spent at work than during an earlier era) also means less time to prepare food. Fast food has thus also grown in popularity in the post-World War II era. Most famously, Ray Kroc and the McDonald brothers found new, rationalized ways to combine factory line assemblage, the convenience of the automobile, and the bounty of industrial agriculture in the provision of hamburgers and fries for the masses (Ritzer, 2011). Since the 1950s, the fast food franchise as a business model has proven profitable and nearly irresistible to Americans who continue to luxuriate in the many wonders of consumer car culture.

Regarding the social structure of American car culture and high caloric fast food, it is worth recalling that neither of these phenomena resulted from preexisting market demand. The US government played a crucial role in creating the context for both our need to drive and desire for Big Macs. Without federal investment in the nation's post-war transportation infrastructure – e.g., the Eisenhower Interstate Highway System – American car culture as we know it would not exist. Government-backed home loans and tax-deductible interest payments on loans were two additional subsidies designed in the 1930s to rev up the moribund economy of the Great Depression by incentivizing people to buy their homes further and further from urban centers.

Meanwhile, government subsidies have been crucial in keeping down the cost of industrial crops. Federally subsidized corn in particular – a source crop in the production of hamburgers (feed for cattle), milkshakes (ditto), and carbonated beverages (high fructose corn syrup) – has done much to help keep fast food dining affordable for everyone (Schlosser, 2012). Thus, our health, something we need to take personal responsibility for throughout our life, is also a social issue, structured and determined by policies paid for through taxes, implemented by our government, and supported vigorously by corporate recipients of federal subsidies.

Can a bureaucratic state protect us from further rationalization?

At their point of origin, government subsidies and the government organizations that administer them were created for the collective benefit of a rapidly growing US population. Weber had a name for these and other institutional forms of rationalization – *bureaucracy*. Though often criticized for being fiscally burdensome and inefficient in its allocation of resources and services, the bureaucratic state – i.e., government – is instrumental in structuring numerous aspects of our lives. For example, our relationships with:

- Each other – what are our obligations to society and how do we collectively manage our affairs through a representative democracy in a capitalist economic system?
- Ourselves – where do we live, where do we work, how do we get there, and how do we nourish ourselves physically, culturally, and intellectually?
- The ecology – what is our experiential access to the natural world, and how do we regulate the extraction of natural resources and pollution of the environment with an eye for the fate of future generations?

Focusing on the last point, there are many examples of the US government stepping in and establishing rules designed to protect the environment and our relationship with it. In August 1916, President Woodrow Wilson signed a legislation establishing the National Park Service which, a hundred years later, manages and maintains 450 recreational, historical, and cultural areas throughout the United States and its territories. Likewise, the decade of the 1970s saw a series of federal laws enacted – the Clean Air Act (1970), the National Environmental Policy Act (1970), the Endangered Species Act (1973), the Clean Water Act (1977), and the Superfund Act (1980), to name a few – that continue to structure our relationship with the environment. As might be expected, there is often pushback from those who stand to profit from the lifting of environmental protections these acts of Congress have put in place. There is reason to believe, however, that further oversight is needed if we are to protect ourselves and the environment from the most egregious outcomes of rationalization focused on extracting profits before the interdependent needs of people and the earth's ecosystems.

The troubled past, present and future of nuclear power generation

Possibly the most astonishing result of rationalization on the part of a bureaucratic state has been the harnessing of atomic power, first for military purposes during WWII, and then as an alternative to fossil fuels for generating electricity. When the Manhattan Project began in 1939, the US government feared the Nazis would soon have the capacity to turn theoretical physics into the ultimate weapon of war, the atomic bomb. Albert Einstein himself co-wrote a letter to then president Franklin D. Roosevelt to warn him precisely of such a possibility (Rhodes, 2012). A team of nuclear physicists led by Robert Oppenheimer were charged by the president through the Army Corps of Engineers to develop an atomic bomb before the Germans did. In July 1945, two months after the Nazis surrendered to the Allied Forces, the first atomic bomb was detonated at the Alamogordo site in southern New Mexico. Weeks later, Japan, which refused unconditional surrender to the

Allies, would be the first and thus far only nation attacked with a nuclear weapon, first in Hiroshima on August 6, and then in Nagasaki on August 9. This gruesome end to World War II – at least 129,000 Japanese people, mainly civilians, met horrible deaths – distills in many ways the confounding relationship the rationalized bureaucratic state and science have with ethical action.

Though nuclear weapons have not been used again in warfare, the production of nuclear warheads accelerated since the end of WWII, a principal element of competition between the United States and the Soviet Union during the Cold War period from the late 1940s through the early 1990s. Even now, there are still some 13,500 detonable nuclear warheads in the world, 6,490 owned by Russia and 5,800 owned by the United States (Kristensen & Korda, 2019a, 2019b). All of these warheads, by the way, are thousands of times more powerful than the two bombs dropped on Japan in 1945. One might reasonably ask, are we maintaining these weapons to make the world a safer, more peaceful place, or the exact opposite? Along with being the poster child for how rationalization can go horribly wrong, the "mutually assured destruction" guaranteed in a hypothetical nuclear war is also a classic example of a *collective action problem*. That is, no reasonable person wants to see the end of the world as we know it, and yet why are we not doing more to prevent this from happening? In a later chapter, we will consider the "contact hypothesis" as a possible solution to the collective action problem of distrust. For now, let's just say that mutually agreed upon rules of conduct (including avoiding a nuclear holocaust) require (a) some degree of mutual trust, and (b) that people actually communicate with each other.

The other side of the nuclear story is, many argue, more benevolent. That is, what if we could take the power inherent in converting matter into energy and use it to reduce our dependence on fossil fuels. Many in the industry have long referred to nuclear power as a "clean" source of energy for powering our electric grid. To the degree that nuclear power plants do not spew out carbon emissions, that much is true. A gram of uranium-235, for example, can produce 1.5 million times the amount of energy contained in a gram of coal (Emsley, 2011). The generation of electricity through nuclear fission, however, comes with its own set of daunting unintended consequences. To be clear, this is not a hypothetical scenario; approximately 20 percent of electricity in the United States comes from nuclear power plants. Though this is a smaller percentage of nuclear power as a source of electricity than that of France (73 percent) or South Korea (29 percent), the United States still generates in absolute terms more electricity from nuclear power plants than any other country on Earth (USEIA, 2019).

The biggest challenge to producing nuclear generated electricity is what to do about the waste. By splitting uranium isotopes, nuclear fission releases energy which in a nuclear reactor is used to boil water to produce steam, which then spins the turbines that generate electricity. The split isotopes, however, are still highly radioactive and may be harmful for tens of thousands of years

(Center for Disease Control and Prevention, 2018). Strange as it may seem, though there are currently 449 nuclear reactors up and running in the world (99 in the United States, alone), there is not yet a clear cut solution for what to do with nuclear waste (Rouse, 2019). For the time being, nuclear waste in the form of spent fuel rods is mainly stored in water tanks on or near the site of nuclear power plants. Currently, the two principal recommendations for addressing this situation are long-term vigilance and deep underground storage.

The first option requires that the human race designate a site that would be contained, protected, and unexposed to the natural environment for possibly 100,000 years. As mentioned above, we as a species have only been around for about 200,000 years and, to give one long-lasting example, the civilization centered around ancient Rome lasted a mere 2,000 years. Bureaucratically, the long-term vigilance of nuclear waste seems unrealistic. The second option, deep subterranean storage, appears to be a more likely alternative. This too, however, has its own set of problems. For decades the federal government in the United States has made efforts to consolidate all waste produced from the nation's nuclear reactors in one centralized storage facility – Yucca Mountain, a deep geological nuclear waste repository in the Nevada desert. This program, however, was defunded by the Obama administration in 2011 after much political opposition from Nevadans and others around the country who did not like the idea of massive shipments of nuclear waste passing through populated areas in railcars or on the trailers of long-haul semi-trucks (Northey, 2011). Of course, keeping spent nuclear fuel rods on site is no less risky, as made tragically apparent at the Japanese Fukushima Power Plant in 2011 when a combination earthquake/tsunami resulted in three nuclear reactors melting down (Rogers, 2011).

Shackled inside the iron cage?

One metaphor from Weber's writing on rationalization seems particularly relevant here: the *iron cage* symbolized for Weber the possibility in a rationalized society that we would create a world of our own making which we really would prefer not to live in (Weber, 2013). Some have argued that this sociological approach to structure (the social, economic, and cultural framework of constraints and opportunities humans have built up over time) and agency (our ability as people to effect change in the world) is pessimistic. It must be acknowledged, however, that, at the very least, Weber perceived the nature of the problem: An escape from the iron cage of industrial capitalism and all its unintended social and environmental consequences will require oceans of coordinated deliberate action.

The road from here to freedom from the iron cage is unclear. This is in part because there are no established maps guiding us to our destination. Also, the road is long, likely lasting generations. That being the case, perhaps a good place to begin considering the possibility of socioecological change is in childhood. It is at that stage that we learn to be who we are, acquiring habits of mind and practice that, for better or worse, stick with us the rest of our lives. Fortunately, many sociologists have put a lot of thought into the dynamics and challenges surrounding childhood socialization, a topic to which we will turn in the following chapter.

Glossary

Atrophy of Vigilance   an idea developed by William Freudenburg and Robert Gramling capturing the declining levels of vigilance among organizations charged with managing environmental hazards.

Bureaucracy   The institutional form of rationalization, consisting of hierarchical levels of organization that are efficient, yet inherently undemocratic. Bureaucracy was critically studied and defined by Max Weber.

Collective Action Problem   A situation where most perceive a common problem in need of addressing, yet lack the collective will to do anything about it.

Energy Slaves   An idea developed by Buckminster Fuller quantifying the energy input required to produce goods and services in modern society in terms of human units of effort; e.g., it would take the effort of five energy slaves to maintain a 100-watt lightbulb fully lit.

Iron Cage   Max Weber's dystopic depiction of rationalization which, in its extreme form, subjugates human individuality, creativity, and enchantment with the natural world to the modern imperatives of efficiency, calculability, and profit.

Profit Maximization   A rational assessment of costs and revenue which prioritizes higher profits over all other considerations.

Rationalization   Society's tendency in modern life to base decisions on rational calculation. An idea defined and popularized by Weber, rationalization is often conceptualized to include three main components – predictability, calculability, and efficiency.

Questions

1   Over the past week, how have rationalized pressures of efficiency, calculability, and profit maximization affected your actions? Did you decide to take these actions yourself, or were they determined by some other person or factor outside your control? Did any unintended consequences result from these actions? Do you think our culture of rationalization works mainly in your favor, or not? Why?

2   Now consider an ideal, value or worldview that matters more to you than profit maximization or efficiency. In what ways is it threatened by the forces of rationalization? What have you done to protect these ideals? In which ways does rationalization perhaps support these ideals?

3   How many bureaucracies have you interacted with in the past 24 hours? What are the benefits you have gained through your relationship with

one of these bureaucracies? What annoys you most about your relationship with this bureaucracy? Who do you think benefits the most from bureaucracies – individuals within society, society as a whole, or the bureaucracy itself? Explain.

4   Two aspects of World War II – the use of nuclear weapons and the Holocaust – are often pointed to as examples of how rationalization can result in extremely irrational outcomes. In our current time, where do you see rationalization going horribly wrong? Max Weber was himself rather pessimistic with regard to the future of industrial capitalism, but what hope do you see in terms of us escaping as a society from the iron cage of our own making?

References

Boase, J., Horrigan, J., Wellman, B., & Rainie, L. (2006). *The strength of internet ties*: Pew Internet and American Life Project.

Bowles, N. (2018). A dark consensus about screens and kids begins to emerge in Silicon Valley. *The New York Times*, October 29, A3 N.

Carrington, D. (2015). Wide range of cars emit more pollution in realistic driving tests, data shows. *The Guardian, 30*.

Center for Disease Control and Prevention. (2018). Radioisotope brief: Uranium. Retrieved from https://www.cdc.gov/nceh/radiation/emergencies/isotopes/uranium.htm

Emsley, J. (2011). *Nature's building blocks: An AZ guide to the elements*: Oxford University Press.

Ewing, J. (2017). *Faster, higher, farther: The inside story of the Volkswagen scandal*: Random House.

Flegal, K. M., Kruszon-Moran, D., Carroll, M. D., Fryar, C. D., & Ogden, C. L. (2016). Trends in obesity among adults in the United States, 2005 to 2014. *JAMA, 315*(21), 2284–2291.

Fuller, R. B. (2001). *Buckminster Fuller: Anthology for the new millennium*: Macmillan.

Kim, S. D. (2017). Characterization of unknown unknowns using separation principles in case study on Deepwater Horizon oil spill. *Journal of Risk Research, 20*(1), 151–168.

Kristensen, H. M., & Korda, M. (2019a). Russian nuclear forces, 2019. *Bulletin of the Atomic Scientists, 75*(2), 73–84.

Kristensen, H. M., & Korda, M. (2019b). United States nuclear forces, 2019. *Bulletin of the Atomic Scientists, 75*(3), 122–134.

Lusk, J. L. (2016). The evolving role of the USDA in the food and agricultural economy – Research summary. (No. 06933). *Mercatus Center at George Mason University*.

Massey, D. S. (2005). *Strangers in a strange land: Humans in an urbanizing world*: WW Norton.

National Highway Traffic Safety Administration. (2019). *Driver Electronic Device Use in 2017* (Traffic Safety Facts Research Note. Report No. DOT HS 812 665). Washington, DC.

Northey, H. (2011). GAO: Death of Yucca Mountain caused by political maneuvering. *New York Times*, May 10. Retrieved from https://archive.nytimes.com/www.nytimes.com/gwire/2011/05/10/10greenwire-gao-death-of-yucca-mountain-caused-by-politica-36298.html. Accessed September 15, 2021.

Panter-Brick, C., Layton, R. H., & Rowley-Conwy, P. (2001). *Hunter-gatherers: An interdisciplinary perspective* (Vol. 13): Cambridge University Press.

Rhodes, R. (2012). *The making of the atomic bomb*: Simon and Schuster.

Ritzer, G. (2011). *The McDonaldization of society 6*: Pine Forge Press.

Robertson, C., & Krauss, C. (2010). Gulf spill is the largest of its kind, scientists say. *The New York Times*, August 3, A14.

Rogers, S. (2011). Nuclear power plant accidents: listed and ranked since 1952. *The Guardian*, March 18. Retrieved from http://www.theguardian.com/news/datablog/2011/mar/14/nuclear-power-plant-accidents-list-rank#data. Accessed September 15, 2021.

Rouse, S. (2019). These countries have the most nuclear reactors. Retrieved from https://www.weforum.org/agenda/2019/11/countries-that-have-the-most-nuclear-power-alternative-energy-electricity-climate-change/

Schlosser, E. (2012). *Fast food nation: The dark side of the all-American meal*: Houghton Mifflin Harcourt.

Schmandt-Besserat, D. (2010). *How writing came about*: University of Texas Press.

Silverstein, S., Freeman, N., & Kennedy, A. P. (1964). *The giving tree*: HarperCollins.

Tierney, K. J., & Quarantelli, E. L. (1992). Social aspects of the Exxon Valdez oil spill. *Industrial Crisis Quarterly, 6*(3), 167–173. doi:10.1177/108602669200600301

USDA. (2019). Ag and food sectors and the economy. *Ag and food statistics: Charting the essentials*. Retrieved from https://www.ers.usda.gov/data-products/ag-and-food-statistics-charting-the-essentials/ag-and-food-sectors-and-the-economy.aspx

USDA. (2020). Farm labor. Retrieved from https://www.ers.usda.gov/topics/farm-economy/farm-labor/

USEIA. (2019). Nuclear explained Nuclear power plants. Retrieved from https://www.eia.gov/energyexplained/nuclear/nuclear-power-plants.php

Von Soden, W. (1994). *The ancient orient: An introduction to the study of the ancient Near East*: Eerdmans Pub Co.

Weber, M. (1978). *Economy and society: An outline of interpretive sociology* (Vol. 1): University of California Press.

Weber, M. (2013). *The Protestant ethic and the spirit of capitalism*: Routledge.

3   The nature of nurture

A lifetime of socialization

Traveling abroad in a non-English speaking country can be both exhilarating and profoundly frustrating. Are you comfortable picking up conversations with strangers, or are you overly self-conscious, concerned about what people think of your foreign ways and speech? Do you enjoy exploring and trying new things, or are you more of a homebody? Whatever your temperament, allowing yourself to enter a culture different from your own is a visceral way of perceiving the social construction of just about everything. At least if you are paying attention. Once you have turned off your mobile devices and left behind the comfort of fellow English-speakers, you may find yourself in a situation where everything you ever learned, especially with regard to language, simply does not apply. In the short run, of course, you can get by with pointing at the map or menu, improvisational sign language, or the cultural privilege of your nation's economic power – i.e., most anywhere you go, someone speaks a little English.

If you are in it for the long haul, however, you probably will want to learn the local language. As a monolingual third-generation Mexican American, learning to speak Spanish as a Peace Corps volunteer in my twenties was a revelation. Though both my parents spoke Spanish as their first language and I took Spanish courses through high school and college, I never felt comfortable speaking it. One thing became apparent to me only years later – exposure matters. Being the youngest cousins (my mother was the last of 13 children), my siblings and I had the least interaction with our Spanish-only-speaking grandparents. I only knew one of my monolingual Mexican grandparents, personally – he died at age 98 when I was five. As for my bilingual parents and other relatives, in practice, there was little incentive to speak Spanish if we both knew we could communicate in English.

There is another bleaker side to this story. When my parents were growing up in the 1930s and 1940s, both of them were prohibited from and could be punished for speaking Spanish in the public schools they attended. That and discriminatory experiences in the housing market and at work, made them leery of having their own children face similar obstacles. Raising us to speak only English seemed like a reasonable parenting strategy at the time.

DOI: 10.4324/9781003110668-3

Thus, language, this thing I had occasionally associated with feelings of guilt for not having learned more than one, was highly dependent on social context. The literature on mastery reveals that 10,000 hours of deliberate, focused practice is a pretty good average for the time it takes your brain to master challenging elements of human culture such as playing with precision and grace a musical instrument, becoming a world-class athlete, and speaking a language with fluency (Ericsson, 2006). With time and dedication and, not incidentally, the institutional support of the US government, I could learn to speak Spanish confidently as an adult, something I was able to do with my parents for the first time when they visited me in Costa Rica in 1993.

I bring up this personal experience with language and foreign travel for a couple of reasons. First, it reminds us that many of the most important things we learn happen when we are small children, usually with very little awareness that we are learning at all. This precious time we have all passed through, essential for everything else that follows, is known in sociology as *childhood socialization*. You only get one shot, and you have virtually no say in what it is you learn at this early age. Second, I want to underscore the fact that, though it is a critical stage, childhood socialization does not determine everything, and in fact there is a mountain of evidence that we can continue to learn and make important changes in our lives and in the world at any age. The key theme undergirding opportunities to learn and change: Social context matters.

In the previous chapter on rationalization, we got a sense for the big picture within industrial capitalism where the dominant cultural tendency towards efficiency, calculability, and predictability in all things has resulted in, along with most of the conveniences and comforts of modern life, some distressing unintended consequences. Within sociology, this is considered a "macro" point of view, a broad theoretical perspective, tracing historical, cultural, social, and economic trends that structure and constrain the way we live our lives, today. When considering socialization, however, our gaze shifts to the "micro" level of social interaction.

Symbolic interaction: the source of self in everybody else

In the early 20th century, George Herbert Mead established his reputation as a sociologist obsessed with the ways in which social reality is constructed at the micro level, through face-to-face encounters. He referred to this dynamic as *symbolic interaction*. In a stark example of life imitating social theory, we know about Mead's ideas, mainly through his interactions with students: He never wrote a book about symbolic interaction but, in person, he was a skilled communicator and engaging lecturer. It was not until after his death that his biggest fans – students at the University of Chicago – amassed their notes and transcribed his lectures into written form.

In his posthumously published book, *Mind, Self and Society* (1934), Mead (i.e., his students) laid out an argument for how we come to recognize ourselves as individuals, distinct from everyone and everything around us. This shift, he informs us, corresponds to the difference between the first-person pronoun "I," and the first-person direct object "me." This is another way of saying that babies and small children, more than just believing they are the center of the universe, believe they are the universe itself – i.e., there is no distinction between themselves as individuals and everything else around them. It is only over time through interaction with others that they begin to step outside themselves and see that "I" (how I perceive everything) is also "me" (how others perceive me). And that, according to Mead, is the origin of the self.

Beyond simply pointing to the stage in life this change occurs, Mead also describes the specific mechanism that makes this possible – language. In doing so, he provides a profoundly sociological explanation for the origin of the self. Most children begin speaking between two and three years of age and, though dependent on having the opportunity to interact with and imitate the words and grammar used by other humans, this development appears to be hardwired in our DNA as humans. The most obvious and perhaps most gratifying thing about this stage as a parent is that you can finally begin to communicate clearly with your child. This is no doubt gratifying for small children as well who can now articulate their needs, complaints, and whatever else is on their mind.

The subtle twist that Mead brings to this scenario is that, not only do children begin to communicate with others at this state, but they also begin to communicate with themselves. That is, the continuous internal conversation we have day-in/day-out that we often refer to as "thoughts," cannot exist without a language (English, presumably, for most of you reading this text) originating in the social world. In the most basic way possible, our sense of self depends on our social interactions with others. In case there are any doubts about the validity of this argument, there are, sadly, cases – often resulting from long-term abuse – where children were not socialized into language at this crucial developmental stage. When this occurs, it is nearly impossible to regain at a later age the lost opportunity for language acquisition and a distinctive, socially based sense of self.

Plays well with others: interdependence as a marker of maturity

Thankfully, for most of us, the emergence of the self in early childhood signals the beginning of an even stronger sense of interdependence with others that continues throughout our life. In that regard, Mead spelled out a three-stage developmental process leading us interactively into responsible adulthood.

The *Preparatory Stage* is largely one of mimicry; from birth to about the age they begin to speak, children do their best to imitate those around them. This is often one of the greatest pleasures of being around infants – at just a few months of age babies can make their parents proud by smiling back at an exaggerated coochy-coo, or some other silly facial expression designed to elicit exactly that response. As toddlers become more vocal, parents can anticipate the arrival of language with the increased tendency to imitate specific sounds and words. There was a time when my daughter was two-years-old and we would drive around with a children's playlist in the car that included a cowboy song she liked quite a bit. Despite what I know about childhood socialization, I was still astonished to hear her cry out, "YEE-hah!" after I served her applesauce for dessert one afternoon.

In the *Play Stage*, according to Mead, imitation is taken up a couple of notches, and children between the ages of two and six begin to mimic the behaviors of those around them. Perhaps at no other time in life is the adage, "Do as I say, not as I do," a more useless proposition. On the positive side, there is no better time in life as a parent to teach by example; that is, children at this stage are practically programmed to imitate the roles and behaviors of those closest to them. Feeding baby dolls, baking a pretend cake, taking a sibling's temperature, or trying to hear someone's heartbeat with a toy stethoscope are all behaviors children base on their keen observations of the adults around them. Of course, parents intent on setting a good example will soon discover there are a myriad of other entities vying for their children's attention. Advertisers, as we will discuss further below, have been especially efficacious in their application of academic research on childhood socialization to further their profit-generating goals.

The final stage of childhood socialization described by Mead is the *Game Stage*. Here, children age seven and older, go beyond simply imitating adult roles and start to contend with peer interactions. Whereas in the Play Stage it is not unusual to see preschoolers engage in "parallel play" – though in the same physical space, they occupy themselves with separate individual activities – children entering grade school are developmentally much more able to participate in interactive, rule-based games. Organized sports (as opposed to chaotic tee-ball, or soccer games where players are unsure about which direction to score a goal) are now a real possibility. Rules provide structure to play, but also call attention to our interdependence with others – we must all agree on the rules for a game to have a meaningful outcome.

It is at this point, Mead informs us, that children begin to develop a sense of the *generalized other*. To get along in this world you need to both understand and learn how to be understood by others within a world tacitly governed by rules of social engagement; adhering to these rules tends to generate good will and a sense of mutual respect. In contrast to the commonsensical view that maturity is best indicated by ever greater levels of independence as we grow into adulthood, Mead's work suggests that true maturity is marked by *interdependence* and the ability to engage respectfully with others on the basis of a common system of rules for interaction.

Though Mead's theories on socialization apply mainly to children, it is worth considering how some of these processes continue throughout one's life. Anyone who has tried to learn to play a musical instrument, try out a new sport, or write a persuasive essay knows that, along with good instruction and practice, mimicry can be an effective learning tool. Moreover, you may have noticed, after spending a period of time with a new group of friends or coworkers, you begin to repeat phrases you heard them say. Regarding role-playing, many adults who have begun a new position at work for which they are qualified but have never before occupied remember feeling early on like an imposter. Sometimes the best strategy in that instance is to "fake it until you make it." In his classic ethnography, *Boys in White* (1961), sociologist Howard Becker and his colleagues describe how young medical students must overcome their anxiety and act confident in front of patients in order to become effective physicians; one telling insight – wearing a white lab coat helps.

Taken out of the context of childhood development, it is clear the Game Stage persists throughout our life, as well. In many ways, to be alive and an adult is to be aware of the rules of social interaction; balancing our need for personal advantage with our responsibilities to those with whom we most frequently interact and the "generalized other" of the broader society. We are not always successful in this last endeavor, but as the previous chapter's conclusion made clear, our survival as a species may depend on our willingness to try.

Socialization through sales and marketing

As alluded to above, the producers of goods and services in a market economy have a vested interest in shaping the arc of our socialization so that it leans towards buying more stuff. In Chapter 2, we saw how fossil fuels have accelerated the rationalized locomotives of production in industry, agriculture, and transportation. The whole thing is liable to run off the rails, however, if no one wants to buy what the producers are pumping out. For this reason, persuasion is often the critical link that holds together the consumer and producer cars on the choo-choo train of industrial capitalism, if you will. Arguably, salesmanship has been with us since the first time there was ever something for sale. However, sometime during the first half of the 20th century, telecommunications technology began to transform sales from a matter of face-to-face convincing to the coaxing of entire populations to desire things they had not actually considered before.

Intellectual interest in the persuasive power of electronic media, however, did not really emerge until the 1930s, and not for the reasons just mentioned. Instead, the rise and democratic election of the Nazi Party into power in 1933 had many – notably, Jewish intellectuals in Germany – trying to understand how one of the most highly educated, technologically advanced nations in the world could turn its government over to fascist rule. A group of social theorists known as the Frankfurt School began to look closely at the role of propaganda in persuading the German population that the mass genocide of

Jews, homosexuals, Gypsies, and other disparaged groups was the "ultimate solution" which would purify the nation and lead it out of economic hardship. Propaganda Minister Joseph Goebbels effectively deployed the instruments of mass media including newsreels and radio to transmit an ideology of hate and racism that turned the majority population against a class deemed genetically inferior and responsible for all that was bad in Germany.

After the war, members of the Frankfurt School who had taken exile in the United States began to take the lessons and theoretical tools learned during Nazi Germany and apply them to the contemporary setting. Who could be sure that what happened in Germany in the 1930s would not happen again? Given the growing role of the United States in the cultural production of film and music, and the emergence of television as a captivating household visual medium it was perhaps worth reconsidering the influence of electronic media in our collective socialization.

Marcuse's synthesis of Freud and Marx

One of the more influential Frankfurt School intellectuals in exile – especially among American countercultural activists of the 1960s – was Herbert Marcuse. Though his early work focused primarily on the writings of Karl Marx, by the time he arrived in the United States, Marcuse had begun a critical synthesis of the works of Marx and the psychoanalytic approach of Sigmund Freud with the end goal of better understanding contemporary consumer society. Before moving forward with Marcuse, it is worth stepping back a bit to consider relevant elements of Freud that helped motivate the former's work on consumer culture.

Near the end of his life, Freud published his last and starkly pessimistic book, *Civilization and Its Discontents* (1989). In it, referencing ancient Greek mythology, he ominously anticipated the defeat of Eros (the god of love) by Thanantos (the god of death) in the not too distant future. Given the ravages of the First World War – not to mention Freud's personal long and painful bout with cancer of the palette – this was perhaps a not unreasonable forecast. In less loftier terms, Freud asserts in *Civilization and Its Discontents* that for civilization to work individuals need to repress their more base animal-like instincts of aggression and sexual desire. That is, some modicum of self-control is necessary for us to make it through the day without the incessant threat of violence and sexual come-ons. Fair enough. However, for Freud there are inherent limits to our ability to repress our instinctual energies. More to the point, if we try to somehow contain the expression of our aggressive, sexual tendencies in one part of our lives, they will likely manifest themselves somewhere else. This could range innocuously from neurotic facial tics, nervous laughter, and "Freudian slips" of the tongue to, more disconcertingly, actual outbreaks of violence.

For Freud, most definitely not a sociologist, the one lingering source of hope was what he called "sublimation" – the redirection of instinctual energy towards more productive and ultimately satisfying ends. Through discipline and practice, one could learn to harness one's base energies to, for example, learn carpentry, pick up ballroom dancing, become really good at backgammon, or even write a career-capping if somewhat depressing book on psychoanalysis. These were the kinds of efforts individuals could make to ward off the looming threat of Thanatos on modern civilization, efforts which, in Freud's work, were notably lacking in any sense of collective purpose or coordinated political action.

Marx, by contrast, was at his core an optimist who anticipated rising political consciousness among the workers of the world which would result in a social revolution liberating them from the shackles of industrial capitalism. For Marcuse, Marx's work was foundational while Freud's work provided a meaningful lens into the psyche of 20th-century modern life. Both their insights, however, would need to be reinterpreted through the post-World War II experience of economic abundance. To begin with, the contrast Marcuse experienced between Depression-era Germany in the 1930s and his life in the 1950s United States where he ultimately remained after the war could not have been greater. The relative opulence of mid-century American life where suburban homes were beginning to brim with tail-finned cars, washing machines, kitchen appliances, and black-and-white television sets made it appear unlikely that a global working-class revolution was right around the corner.

Contrary to Marx's expectations, workers were beginning to reap directly the benefits of capitalist production. This same abundance, according to Marcuse, also signaled a rethinking of Freudian repression. Given the many needs and desires that could now be satisfied through capitalist abundance, repression might no longer be necessary (Marcuse, 2012). Eros, as Marcuse and many of his countercultural admirers concluded, was now on the rise. Love was maybe not all you needed to change the world, but the New Left appeared to be shifting its political focus away from the heroic Marxist working class towards a broader array of cultural, identity, and ecological concerns.

Counterculture shackled by "repressive de-sublimation"

With the publication of *Eros and Civilization* in 1955, Marcuse – sometimes referred to as the "father of the New Left" – helped open the door to the countercultural political wave of the 1960s. In it, though acknowledging Freud's insight that civilization required some degree of self-control over instinctual energies, Marcuse argued that, in the age of abundance, we were all suffering from what he called "surplus repression." That is, though the first half of the 20th century left the global population with a strong sense of restraint and frugality in the face of economic scarcity, postwar abundance, by contrast, should provide an historic opportunity to shed the chains of psychic repression. According to Marcuse, our instinctual energies – especially the sexual ones – could now be more freely expressed through the liberating process of *de-sublimation*. Along with shifting social norms, technological changes such as the widespread availability of the automobile and, beginning in the 1960s, the commercial availability of an oral contraceptive, known simply as "The Pill," greatly facilitated access to less restrained lifestyles among average Americans compared to what had come before.

These new freedoms, however, did not come without their own risks. Beyond the increased epidemiological risks of sexually transmitted diseases, Marcuse was himself much more concerned about the ways capitalism might tap into this new age of de-sublimation, essentially harnessing repressed sexual and aggressive energies for profit. This was not, by the way, just a hypothetical proposition. Much has been written and at least one successful television series has been based on the influence of psychoanalysis on advertisers in the second half of the 20th century (Curtis, Kelsall, & Curtis, 2002). Marketing executives such as Edward Bernays (Freud's nephew) and Ernst Dichter utilized insights from psychoanalysis to play on people's sexual insecurities and the indeterminate social roles of men and women in an increasingly commodified environment.

In one classic example, focus group sessions – a now widely used qualitative research method first developed by Dichter for marketing research – revealed that women in the 1950s were deeply ambivalent about using Betty Crocker instant cake mix. In these group interviews, women admitted they felt they were not contributing enough in the creation of this mostly prefabricated cake. The solution, Dichter decided was to require that women contribute a single egg to the mix before putting it in the oven. Perhaps the allusion to sexual reproduction was coincidental, perhaps it was not. Whatever the case may be, sales of Betty Crocker instant cake mix took off after this minor tweak in the recipe (Curtis et al., 2002).

A more obvious example of capitalism tapping into repressed sexual desires was the monumental success of *Playboy* magazine, famously launched in 1953 with an issue featuring a nude centerfold of Marilyn Monroe. Its publisher Hugh Hefner had correctly calculated that there was a massive market among repressed American men for his publication which, as Marcuse saw it, engaged in *repressive desublimation*. That is, instead of redirecting our instinctual energies towards creative expression and the healthy release of sexual desire, publications such as *Playboy* snuffed the liberating potential of desublimation, objectifying women and commodifying sex, in the process. There is an ironic twist to this story – in 1970, after attaining notoriety as a countercultural philosopher-guru, *Playboy* requested a feature interview with the 70-year-old Marcuse. He agreed to the interview under one condition – Marcuse would also be that month's *Playboy* centerfold (Pollard, 2017). In one clever gesture, Marcuse clearly indicated the flimsy intellectual fig leaf *Playboy* often used to obscure its central project of exchanging porn for cash. Not surprisingly, Hefner did not accept Marcuse's proposal.

Even if you cannot swallow whole his theoretical synthesis of Freud and Marx, it is worth acknowledging Marcuse's central insight: There are paid professionals hired by industry who have as their primary goal – utilizing all the means of media, technology, and academic research at their disposal – the manipulation of human weaknesses and emotional vulnerabilities for the extraction of profit. Moreover, according to Marcuse, with each successful manipulation we become a little less human as our desire for meaningful interaction and social communion with others is replaced by the mindless consumption of goods and entertainment.

Screen-based socialization

Today, seven decades after the publication of *Eros and Civilization*, attempts by marketers at grabbing our attention are all the more ubiquitous and, for many, unsettling. Not surprisingly, researchers and parents among others have begun to notice the disproportionate amount of time children spend in front of screens. As mentioned in the previous chapter, there is some evidence that electronic devices serve mainly to reinforce our preexisting dispositions – extroverts can attend even more festive soirées filled with joy and laughter than ever before, while introverts may find themselves spending a lot more time alone in their room. The dispositions of children and adolescents, however, are not quite yet gelled. What are the consequences of a childhood filled with hours, day-in/day-out, year after year looking at a screen?

For a while it seemed we could comfort ourselves with the notion that the "jury is still out" on this topic. Recently, however, a number of studies have discovered some remarkable associations between the social lives of teens and the use of electronic media. A research paper published by the children's advocacy group Common Sense Media (CSM, 2015) found that US teens used on average nine hours of electronic media per day. Looking a little more closely at the data, boys spent on average much more time per day playing video games than girls (56 versus 7 minutes), and girls on average spent more time on social media than boys (92 versus 52 minutes). CSM also found evidence in their data for the digital divide – i.e., unequal access to electronic media on the basis of class and race. African American teens, for example, have less access to computers and smartphones than whites, but when they do have access they spend more time using them on average than either whites or Latinos.

On a somewhat more disturbing note, Psychologist Jean Twenge has found using longitudinal survey data that the advent of the iPhone in 2007, tracks closely with a spike in teens feeling lonely and not getting a full night's sleep (Twenge, 2017). Also, teens in the iPhone age are notably less likely to go out socially without their parents, or go on a date then they were prior to 2007. There are of course some positive takes on this, as well. Seniors in high school, for example, are about 15 percent less likely to have a driver's license in 2015 than in 1976, an ongoing trend that accelerated after the near market saturation of smartphones over the past decade. Given the fact that motor vehicle accidents are and have been for some time the leading cause of death among US teens, putting driving off for a few extra years may not be a bad idea. Moreover, a greater percentage of young adults using alternative forms of transportation as a trend is, from an environmental perspective, a positive development, as well.

One of the more intriguing findings in this area of research comes from the National Institute on Drug Abuse which found in 2017 a 40-year low in the consumption of illicit drugs among American teens (Johnston, et al., 2017). Though a causal connection has not yet been established, a few researchers are positing that teens are increasingly satisfying their stage-in-life needs for thrill-seeking and novelty through electronic media, and less so through controlled substances.

What comes first, the medium or the message?

Socialization into adulthood, it would appear, is a very different experience from what George Herbert Mead described to his undergraduate students at the University of Chicago a century ago. Even the term socialization seems not quite accurate when so much time in a day is spent in front of a screen – a 2017 study by the Pew Research Center found that over a quarter of American adults say that they are "almost constantly" online (Perrin & Kumar, 2019). How can we better understand the way new media and technology is shaping who we are as a society?

In 1964, Canadian philosopher Marshall McLuhan famously observed that "the medium is the message," meaning the way information was being conveyed – at that time television was the main focus of attention – was strongly shaping the message itself. The first televised presidential debate in 1960 between the relatively young and strapping John F. Kennedy and the slightly disheveled, not-quite-recently shaven Richard Nixon is an iconic example. Polls found that people who listened to the debate on radio thought Nixon had won; while those who witnessed it on television saw Kennedy as the clear victor (Druckman, 2003). Today, serious concerns have been raised about the state of our democracy in the face of "fake news" and the "echo chamber" that result from surrounding ourselves with only opinions with which we agree. Now more than ever, the online medium itself appears to be shaping our perception of reality.

Before we get carried away, however, it is worth stepping back and considering the message itself. Who formulated it to begin with? What are their interests in having this message received? How have stereotypes and chauvinism been deployed in the message to make it more palatable to particular audiences whose support is being sought for political gain? In this last regard, sociology has a lot to tell us about how, along with language acquisition, proper comportment at school, and the rules of baseball, we are also socialized into a legacy of bias and prejudice handed down to us from those who came before. More often than not, these prejudices function to maintain a veneer that everything – including the unequal treatment of certain groups in society – is all right, and provide a folk justification for why some people have or should have a higher social rank than others.

The built-in biases of gender socialization

Beliefs about the inherent differences between men and women provide a highly relevant example of how socialization can work to normalize ranked social roles, in this case on the basis of gender. The thing about being socialized into something from an early age is it can feel like you never actually had to learn it; as though things like social roles and norms have always been that way, and should not be called into question. You might say they just feel natural. Even within academia there are researchers who have forwarded the argument that gender differences, specifically the social differences between men and women and their respective roles in society, are largely built into our DNA. Evolutionary psychologist, David Buss, for example, has long argued that the tendency – as he sees it – for women to desire a single strong, loyal, and intelligent provider as a mate, and men to desire multiple, beautiful, and faithful females represents an evolutionary outcome based on millennia of successful mating strategies within our species (Buss, 2016).

While not denying that there exist systematic physical differences between men and women within the human population, sociologists are much more skeptical – suspicious, even – of primarily biological explanations for cultural and social behavioral differences. The vocabulary sociologists use to make the distinction here is clear: *sex* refers to our physiology, the physical, anatomical differences between men and women; while *gender* refers to the cultural, socialized differences between men and women that we learn in early childhood and which are reinforced throughout our lives. We are born with a sex, but we learn and, in some sense, learn to perform gender as actors on the social stage of life.

Gender transformation: some recent examples

In response to biological explanations for human behavior such as those posited by Buss, sociologists might point to recent changes in gender roles that have had enormous significance for the relative power of men versus women in modern society and could not possibly have happened on the basis of evolutionary advantage.

Economic necessity, for one, has led to a number of remarkable redefinitions of gender roles during the 20th century. Between 1941 and 1945, the United States faced a massive labor shortage as young men joined the ranks of the military to fight in World War II. At the same time prime-aged workers were being dispatched overseas, production in agriculture and industry to support the war effort needed to be ramped up. Faced with a crisis, the US government took two large-scale role-changing measures. First, it sent agents to Mexico to recruit laborers desperately needed on American farms. Second, and more directly tied to the question of shifting gender roles, the US government began a national campaign to recruit women to work in industry, and in doing so, essentially redefined male-dominated occupations as available and appropriate for women.

In addition to providing skills training in welding and the use of heavy machinery, a good deal of cultural retraining around gender was required, as well. The latter was carried out mainly through the use of posters, newsreels, and other propaganda alluding to the similarities between using an electric mixer in the kitchen and a power drill, for example, or cutting a pattern for a dress and cutting sheet metal for planes. All of this seems vaguely ridiculous now, but gender-shifting propaganda – including the now iconic figure of Rosie the Riveter – was essential for getting broad public buy-in to the war effort (Honey, 1984). Unfortunately, the sociopolitical origin of this shift was made crystal clear near the end of the war when newsreels and magazine articles began to reverse course, reminding mothers of their proper role in the home, and emphasizing the priority for jobs that should be given back to "our boys" returning from combat. Millions of women who had gained valuable skills in the shipyard and factory floor now had to relinquish this work – and their higher wages – for the sake of restoring the traditional rank order of gender in American society.

A much longer standing shift in job opportunities for women would have to wait another 25 years. After two decades of solid growth in post-war America, the US economy began to run into a few significant bumps in the 1970s, notably a couple of major spikes in the price of petroleum. Exposure to increasingly competitive global markets in industry probably did not help matters much, either. Whatever the specific source of the economic downturn, middle- and working-class households who had prospered in the 1950s and 1960s were now finding it difficult to make ends meet on a single income. Women – especially working- and middle-class white women; women of color already had relatively high rates of workforce participation – began to enter the market for jobs en masse. Though this clearly signaled for women a higher level of economic power vis-à-vis men, the timing of this shift in workforce participation strongly suggests economic necessity at the household level was its key determinant.

Other examples of shifts in gender roles abound in the 20th century, in many cases not motivated directly by economic transition or trends in the labor market. As hard as it may be to believe, the electrification of homes and the invention of both the airplane and automobile preceded the granting of the right of women to vote in the United States in 1920. Owing little to economic necessity, the success of the Women's Suffrage Movement came about only after decades of political strategizing and struggle against a dominant class of men who felt women were unfit for democratic participation.

A more recent victory for women's rights – specifically for the control they have over their own bodies – took place in 1973 when the US Supreme Court decided in its decision on *Roe v. Wade* that women could legally have an abortion during the first trimester of pregnancy. About a decade before *Roe v. Wade*, the commercial availability of oral contraception which, in addition to the broader cultural impact it had on attitudes towards sexual mores, as mentioned above, gave women an additional level of power and control over sexual reproduction, specifically, when to have children, or whether to have children at all.

Research on the social construction of gender reminds us that aspects of human behavior we may believe are simply part of human nature – our interactions with others, the work we choose, sexual reproduction, the relative status of men and women – often originate in very specific social and historical circumstances. The fact that some of these changes have been brought about through the coordinated efforts of politically minded people in a democracy should be a source of hope. That which we are socialized into may – especially if a critical mass of people believe it is possible and are willing to do something about it – evolve and change in a positive direction.

Shifts in the public's acceptance of non-normative sexuality

A recent example of a coordinated political effort bringing about change in federal law occurred in 2015. That year, the US Supreme Court determined in *Obergefell v. Hodges* that state-level bans on same-sex marriage were unconstitutional. From the outside, the shift from the Defense of Marriage Act passed in Congress in 1996 for the explicit purpose of allowing state-level bans on same-sex marriage, may have seemed like rapid social change. However, taking a longer view, the lesbian, gay, bisexual, transgender, and queer (LGBTQ) movement dating back to the late 1960s had long been making inroads into mainstream culture. Along with signaling a personal liberation from what can be a vindictive heteronormative society, being "out and proud" at the individual level also means it is more likely people everywhere must acknowledge the existence of LGBTQ people in their community, at work, and perhaps most significantly, in their family. This has not been an easy transition, but longitudinal survey research shows a clear though gradual change in attitudes towards the broad acceptance of same-sex relationships, a tendency especially notable among people born after 1982 (Smith, 2011).

The impact of the LGBTQ movement has even shaped the way we think about early childhood socialization. Sociologist Elizabeth Rahilly borrows from French philosopher Michel Foucault when she argues that the parents of "gender-variant" children – i.e., those who identify as a different gender than what they were assigned at birth – often face an ideological "truth regime"; a dominant gender standard that seeks to categorize children into the strict binary of boy or girl (Rahilly, 2015). Using qualitative interview data with the parents of 16 non-gender-normative children, Rahilly finds that even politically liberal parents struggle to raise children outside societal expectations of the gender binary.

In negotiating the truth regime of the gender binary, the parents Rahilly interviewed utilized a series of strategies to help socialize and integrate their gender-variant children. First, *gender hedging* – allowing kids to cross-dress at home, for example, but not at school – is an early effort parents often make so children can be themselves at home in the family environment, but also not risk disapproval in public. With time, as it becomes apparent that their children are not simply going through a passing phase, parents also may engage in *gender literacy*, an effort to understand on a more intellectual basis the distinction between gender and sex, and the fact that these two aspects do not always necessarily align with each other. This is also a time when parents realize that they are not alone and that they can reach out to other parents facing similar challenges.

Lastly, Rahilly describes what she calls *head games* or *playing along*. Here the focus shifts to interactions with others in public, where both parent and child are aware of the uncertain reception of the child's gender ambiguity. At this point, the parent often has to make a choice – being out about the child's gender identity, which may involve some explanation to a non-receptive audience; or just letting a misidentification of a child's gender slide, thus avoiding a difficult or time-consuming interaction, altogether.

Raising any child has never been an easy task, but trying to raise a gender-variant child within the dominant gender binary presents some unique challenges. As Rahilly's research makes clear, however, LGBTQ activism has helped to create a cultural context where gender literacy is possible and parents now have at their disposal a set of tools and strategies, and even a degree of social support, which can all facilitate the process of childhood socialization during uncertain times.

Unequal outcomes and gender discrimination

To be clear, a theoretical basis for identifying examples of the social construction of gender does not equate to a solution for the persistence of gender inequality and discrimination. A report of the National Transgender Discrimination Survey (Grant, Motter, & Tanis, 2011) found that rates of unemployment, housing and work discrimination, and refusal of medical treatment were all higher among transgender respondents than the population average. These levels of unequal treatment were particularly concentrated among transgender individuals of color. Because of their non-normative gender identities, these groups are also much more likely than other groups in society to be the target of harassment and violence. More broadly, LGBTQ youth – more likely to be rejected by their families – are twice as likely to experience homelessness (Morton, Dworsky, & Samuels, 2017), and five times as likely to attempt suicide than their age-equivalent heteronormative peers (Kann, 2011).

As the #MeToo movement has underscored in recent years, women as a category are also much more likely to experience verbal and physical harassment at work on the basis of gender than are men. Lamentably, unequal treatment at the interpersonal level is mirrored at the social structural level of income and occupational status. Women continue to earn on average, and holding constant multiple factors such as education, work experience, specialization, and hours worked per year, about 80 cents to every dollar earned by men (Fontenot, Semega, & Kollar, 2018). There is also evidence that those occupations in which women are most heavily concentrated consistently pay less than those where men are concentrated. This phenomenon known in sociology as "sex segregation" is evident when we see that, though women now make up 32 percent of all lawyers and 34 percent of physicians in the United States, they also constitute 82 percent of the country's social workers and elementary and middle school teachers (Hegewisch & Hartmann, 2014). Moreover, despite decades of being in the workforce at near equal rates as men, women are still more likely to take disproportionate responsibility for raising children, as well as, among siblings, care for aging parents. As a society, we have a long way to go towards ending gender bias and discrimination. That said, major changes in the roles and opportunities available to women that have taken place during the 20th century – often after a long period of political struggle – provide evidence that change is possible.

Ecological feminism

If research on gender shows that social roles and even the process of socialization itself can adjust to new circumstances, in what other parts of modern life can we find evidence for transitions that could lead to positive, more equitable and sustainable outcomes? On a theoretical plane, *ecological feminism*, sometimes referred to simply as ecofeminism, may have some insights. Emerging out of the both the feminist and environmental movements of the 1970s, ecological feminism is a critique of male-dominated industrialism and the exploitation of humans and the natural world for capitalist profit. It shares with feminism a common understanding of *patriarchy* – the unequal exploitive relationship between men and women. Similar to the exploitation of workers by capitalists within Marxism, through patriarchy, men benefit from a social system which devalues the labor of women to the benefit of men. The fact that men and women often occupy the same household underscores the potential for exploitation, abuse, and violence between these two unequal categories. Ecological feminism expands this conceptualization of patriarchy to include the decimation of nature.

Ecofeminism also draws attention to our own alienation from a direct understanding of the natural world. Living within capitalism to some degree requires us all to accept the dominant, often abstract terms of capitalist economics wherein the underlying value in the world is market value. In doing that, we discount our direct knowledge of nature and our relationships with other people. In pre-capitalist societies, women were often the transmitters of traditional knowledge including the use of medicinal plants, foraging, and food preparation (Salleh, 2003). With the arrival of modern medicine and the industrial production of goods, traditional feminine roles such as midwife and herbalist became devalued and disparaged, even, as witches; i.e., women who should not be entrusted with your health and well-being. There is, it should be acknowledged, a strain of romantic thinking within ecofeminism which essentializes women in traditional caretaking roles. The broader point, however, is that lay, everyday knowledge of our interdependence with nature has all but disappeared from modern life.

A key idea elaborated on and utilized within ecological feminism is reproduction. Reproduction from a Marxist perspective refers to those things provided at no costs to the capitalist – child-raising, language acquisition, education, the biological reproduction of more workers – but which are essential to capitalist production. Though housework is itself often tedious and time–consuming, some ecological feminists see within it the potential for positive change, grounded in a direct, harmonious connection to nature. Instead of the competitive disruptions characteristic of capitalism, ecological feminists often stress the nurturing side of home culture which seeks continuity, preservation, and a desire to minimalize the use of natural resources with the knowledge that overexploitation will only undermine the stock of future provisions. This kind of knowledge "holding" is hard-earned, resulting from generations of interactions with the natural world and an intimate understanding of ecosystem limitations (Salleh, 2003).

Reproduction in this context implies a kind of continuity, as in reproducing what has come before, including humans' relationship with the natural world. Production, by contrast, is about extracting material resources and human labor, stripped of context and embedded relationships, for the sole purpose of profit-making. Ecological feminism thus valorizes the "moral economies" of those who still maintain some aspect of their household culture outside of the capitalist sway of market value (Mies, 2014). Whereas Marxism implies that we can never retreat from modernity and the productive innovation of industrial capitalism, ecological feminism reminds us that it is worth holding onto those elements of shared experience with others and the natural world that make us human. For most of us, attaining this ideal would require a resocialization towards reduced reliance on consumer goods and increased engagement with the natural world.

In case one might have quibbles about tinkering with the process of socialization for a more sustainable future, it should be remembered that corporate marketers have had no such reservations, and have lobbied successfully for decades for the deregulation of advertising in their favor, especially that aimed at younger audiences. As sociologist Juliet Schor has argued in her book, *Born to Buy* (2004), marketing researchers have long applied academic research in child psychology and the social sciences to better sell their products. Concepts such as "age compression" – using imagery associated with older children to sell products to younger children, reflected in the term "tween" – and the "nag factor" – taking advantage of less authoritarian parenting styles and the sense of guilt many time-strapped parents feel around child-rearing – have been effectively deployed to socialize children into avid consumers and to increase sales.

Socialization for a more equitable and sustainable future

Given this aggressive media context, it would seem negligent NOT to reconsider how our children are being socialized into adulthood. In this regard, there have been some encouraging developments in recent years. In his book, *Last Child in the Woods,* journalist Richard Louv (2008) describes *nature deficit disorder* as the increasing tendency for children to spend time inside, attached to electronic media and away from nature. He links this trend to concurrent higher rates of childhood obesity, attention deficit disorder, and bullying in American society. This and related academic research have led to a greater awareness of the central importance of being in nature and active for both children and adults. In the next chapter, we will examine a social context that has worked to resocialize people towards nature and a more direct relationship with food production in local agriculture. More broadly, we will consider how certain kinds of social connections are especially good at generating trust and solidarity among groups in society seeking social change.

Glossary

Age Compression   A marketing tactic used to appeal to young audiences by utilizing actors and situations a step or two above the target market's age category.

Childhood Socialization   The sometimes deliberate, sometimes unconscious process of transferring norms, beliefs, and values from one generation to the next.

Frankfurt School   A group of social theorists who began to study the complex intersection of civil society, nationality, ethnicity, and authoritarianism in the early to mid-20th century. Herbert Marcuse and Theodor Adorno were among this group's most influential thinkers.

Game Stage   The last stage of George Herbert Mead's formula for childhood socialization wherein children seven years and older are able to cooperate with peers and follow agreed upon rules of games.

Gender   Cultural, socialized identities, and expectations of femininity and masculinity.

Gender Hedging   Negotiating social constructs of gender and gender boundaries to create more authentic versions of identity that do not necessarily obey dominant social constructions of gender norms; a strategy often taken by the parents of gender-variant children.

Gender Literacy   An effort made by friends or families of gender-variant children to understand the social universe of gender as fact and science.

Moral Economies   An aspect of ecological feminism reflecting an ideal household economy, based on cultural traditions which emphasize continuity, preservation, and the conservation of natural resources.

Nag Factor   A marketing tactic aimed at children which takes advantage of less authoritarian parenting styles.

Patriarchy   A social, cultural, and economic system which benefits men to the detriment of women.

Play Stage   The second stage of Mead's model of childhood socialization wherein children (ages 2–6) utilize mimicry to learn from their parents and caretakers.

Preparatory Stage   The first stage of Mead's model of childhood socialization wherein infants under the age of 2 have not yet differentiated themselves from others and the world around them.

Repressive de-sublimation   Herbert Marcuse's conceptualization of the way marketing and advertising redirects our instinctual creative, sexual, and aggressive impulses towards the purchase of consumer products.

Reproduction   A Marxist conceptualization of the ways in which people are born, raised, and socialized into their roles as workers at no cost to the capitalist in terms of child-rearing, education, language acquisition, etc.

Sex   Biological differences marking male or female characteristics.

Sex Segregation   Gendered patterns in the workforce such that those occupations where women concentrate – e.g., caretakers, administrative assistants, and social workers – on average pay less than those occupations where men concentrate – e.g., firefighters, pilots, and lawyers.

Surplus Repression   Marcuse's assertion that during the relatively prosperous period after WWII, people unnecessarily hung on to social norms of frugality, conformity, and reserve which they had been socialized into during an earlier period of relative austerity.

Symbolic Interaction   A social theory developed by George Herbert Mead which saw the social world as being constructed by micro-level encounters between individuals dependent on a reflexive understanding of how other people perceive them.

Truth Regime   A dominant standard or social norm which attaches morally imperative social expectations to specific categories of people, especially along the lines of gender.

Questions

1   Though our parents and caretakers are usually the ones who determine how we are socialized early on, the older we get the more control we have over what we want to learn. Name one skill you are most proud of having learned deliberately through your own hard work. What strategies did you take to acquire this skill? Did the 10,000-hour rule apply? How much did you depend on others – mentors, teachers, friends, or family – to pick this up?

2   What do you think of Jean Twenge's 2017 study on smartphone usage among US teens? Is this group of young people more lonely, less social, and insomniac then previous generations because of more time spent in front of screens? Also, do you agree with the hypothesis by some researchers that a 40-year low in the use of illicit drugs by teens can be tied

back to the popularity of smartphones, as well? What other alternative explanations do you think are there for the findings in both Twenge and the National Institute on Drug Abuse's studies?

3   In response to sociobiological explanations for presumed difference between the sexes, sociologists point to the ways gender roles and the unequal treatment of women in society are clearly grounded in relationships of power and specific moments in history where shifts in the labor force, economic necessity, and/or political activism have played determining roles in the relative status of men versus women. That is to say, gender is a social construction. Ecological feminism, by contrast, tends to emphasize essential differences between men and women, with women in particular, seen as being more closely bound to nature and thus the obvious purveyors of old traditions and moral economies originating outside the capitalist system. Can these two feminist perspectives ever be reconciled? Should they be?

4   How can we be socialized as humans to bring about a more sustainable future? What types of social institutions do you think will be required to reduce our collective ecological footprint on the planet? Have you ever participated in CSA or lending library in your hometown? What other types of institutions – be it physical buildings, organizations, or simply shared patterns of behavior – could make a difference and reduce our negative impacts on the planet by increasing our positive interactions with each other?

References

Becker, H. S., Geer, B., Hughes, E. C., & Strauss, A. L. (1961). *Boys in white: Student culture in medical school*: Transaction Books.

Buss, D. M. (2016). *Evolution of desire, The*: Springer.

Common Sense Media. (2015). *The common sense census: Media use by tweens and teens*: British Columbia Teachers' Federation.

Curtis, A., Kelsall, L., & Curtis, A. (2002). The century of the self [Documentary film]. *BBC Four*.

Druckman, J. N. (2003). The power of television images: The first Kennedy-Nixon debate revisited. *The Journal of Politics, 65*(2), 559–571.

Ericsson, K. A. (2006). The influence of experience and deliberate practice on the development of superior expert performance. *The Cambridge Handbook of Expertise and Expert Performance, 38*, 685–705.

Fontenot, K., Semega, J., & Kollar, M. (2018). *Income and poverty in the United States: 2017*: US Government Printing Office.

Freud, S. (1989). *Civilization and its discontents*: WW Norton & Company.

Grant, J. M., Motter, L. A., & Tanis, J. (2011). *Injustice at every turn: A report of the national transgender discrimination survey*: National Center for Transgender Equality and National Gay and Lesbian Task Force.

Hegewisch, A., & Hartmann, H. (2014). *Occupational segregation and the gender wage gap: A job half done*: Institute for Women's Policy Research.

Honey, M. (1984). *Creating Rosie the Riveter: Class, gender, and propaganda during World War II*: University of Massachusetts Press.

Johnston, L. D., O'Malley, P. M., Miech, R. A., Bachman, J. G., & Schulenberg, J. E. (2017). *Monitoring the future national survey results on drug use, 1975–2016: Overview, key findings on adolescent drug use*: Institute for Social Research.

Kann, L. (2011). *Sexual identity, sex of sexual contacts, and health-risk behaviors among students in grades 9–12: Youth risk behavior surveillance, selected sites, United States, 2001–2009* (Vol. 60): DIANE Publishing.

Louv, R. (2008). *Last child in the woods: Saving our children from nature-deficit disorder*: Algonquin books.

Marcuse, H. (2012). *Eros and civilization*: Routledge.

Mead, G. H. (1934). *Mind, self and society* (Vol. 111): Chicago University of Chicago Press.

Mies, M. (2014). *Patriarchy and accumulation on a world scale: Women in the international division of labour*: Zed Books Ltd.

Morton, M. H., Dworsky, A., & Samuels, G. M. (2017). *Missed opportunities: Youth homelessness in America. National estimates*: Chapin Hall at the University of Chicago. http://voicesofyouthcount. org/wpcontent/uploads/2017/11/VoYC-National-Estimates-Brief-Chapin-Hall-2017.Pdf.

Perrin, A., & Kumar, M. (2019). About three-in-ten US adults say they are 'almost constantly' online. *Pew Research Center*. https://www.pewresearch.org/fact-tank/2019/07/25/americans-going-online-almost-constantly. Accessed September 15, 2021.

Pollard, C. (2017). The philosopher who was too hot for Playboy. *The Conversation*, October 3. Retrieved from https://theconversation.com/the-philosopher-who-was-too-hot-for-playboy-85002. Accessed September 15, 2021.

Rahilly, E. P. (2015). The gender binary meets the gender-variant child: Parents' negotiations with childhood gender variance. *Gender & Society, 29*(3), 338–361.

Salleh, A. (2003). Ecofeminism as sociology. *Capitalism Nature Socialism, 14*(1), 61–74.

Schor, J. (2004). *Born to buy: The commercialized child and the new consumer culture*: Simon and Schuster.

Smith, T. W. (2011). *Public attitudes toward homosexuality*: NORC at the University of Chicago.

Twenge, J. M. (2017). *IGen: Why today's super-connected kids are growing up less rebellious, more tolerant, less happy--and completely unprepared for adulthood--and what that means for the rest of us*: Simon and Schuster.

4 Local agriculture and the multiplex origins of socioecological change

Like many Americans, the natural surroundings I find myself in now are quite different from the ones I was socialized into as a child. The Sonoran Desert encompassing Phoenix, Arizona in the 1970s had a stark, arid beauty punctuated by saguaro cactuses, rocky vistas, and a never-ending succession of orange and pink sunsets. In the summer months, as I delivered the *Phoenix Gazette* astride my banana seat bicycle, it was also extremely hot.

By contrast, the New England town of Burlington, Vermont where I now work and live experiences a full range of seasonal changes. To wit – humid summer nights cursed by mosquitoes, consecrated by fireflies; the early signs of fall foliage, heralding the compulsory consumption of apple cider and pumpkin flavored products everywhere; real snow in winter; and the miraculous resurrection from the dead of outdoor life in spring. Of course, having grown up in the national media environment of the United States, I was made aware of these things through television, books, and grade school lessons on the "seasons of the year." Sadly, the closest thing I saw to a representation of my own region at the time were the backdrops in Wile E. Coyote and Road Runner animated cartoons.

It was thus with an outsider's sense of curiosity that I began research on local agriculture in Burlington not long after my arrival here in the mid-2000s (Macias, 2008). Why, I thought to myself, are these people so obsessed with local food and farmers markets? From a more sociological perspective, I also wondered what relation does the production and consumption of local agriculture have with the ways people are connected to each other socially? For the first question, I had my suspicions – in 2005, not long after Hurricane Katrina struck the Gulf coast, there was a sudden spike in the cost of petroleum. For the first time since the 1970s, many researchers, journalists, and members of the general public were starting to consider the crucial role hydrocarbons play in on our lives. As highlighted in the earlier chapter on rationalization, we have a broad dependence on fossil fuels for all forms of transportation. Though the connection between the price of petroleum and agricultural production may be less obvious to the average consumer, in 2008, a series of food crises in over 30 countries – notably Egypt, Haiti, Indonesia, and Mexico – made clear the direct impact that higher costs of wheat, rice, and corn production can have on households around the world.

DOI: 10.4324/9781003110668-4

In part as a response to growing awareness, many consumers and some producers in the United States were beginning to turn to local production as a possible alternative to industrial agriculture, now seen as overly reliant on the volatile price of petroleum. Broader claims were being made, however, including in the social sciences. Some researchers argued that, in addition to cutting back on "food miles" – i.e., the distance your food had to travel to arrive at your grocery store – local food was also said to bring local communities closer together, forging bonds of trust and friendship that otherwise might not exist.

Three questions

Can local, sustainable food production be equitable, as well?

For this research, I was interested in three sociological questions. First, was the growing market of local organic food that emerged after 2005 food equitable? That is, did people of different socioeconomic backgrounds have similar levels of access to this market? This is connected to the broader question of "food justice," a movement grounded in the fact that in many regions around the country, lower income families often have little or impeded access to a good selection of healthy food. The consumption of highly processed food which provides much in the way of calories and little in the way of nutrition has been consistently tied to obesity, diabetes, and heart disease. Access to a healthy diet, along with exercise and an active social life, is seen as a key element in living a long and happy life with a lower risk of chronic illness. Unfortunately, as things stand now, on average local organic produce still costs more for the consumer than the alternative. The investment of time – going to a local market, cutting up fruits and vegetables, finding a recipe that works – also presents a major barrier to the consumption of healthy food.

Does local food bring people together or tear them apart?

Aside from the quality of and access to food, Emile Durkheim's views inspired me to explore whether or not local food networks bring people together. Durkheim's theory of social integration focuses on the social context of individuals who left behind traditional life in small rural towns, and seeks to understand how these people would be able to connect with others in a new urban environment. Over a hundred years later, this is still an open question – how do people in modern life feel at all connected to the people around them? For his part Durkheim felt local volunteer organizations would help take the place of churches and town halls where people had come together in the past. It is with this lens that I also approach the question of local agriculture: Do farmers markets, community-supported agriculture (CSAs), and community gardens bring people together? Or, do they have the exact opposite effect? More to the point, might local agriculture actually increase divisions in a community, drawing a line between people who have the time and resources to participate in it, and those who do not? Is local agriculture not just another form of conspicuous consumption whose ultimate function is to delineate the haves from the have-nots?

Do local forms of agriculture generate "natural human capital"?

Lastly, I wanted to explore directly in my own region the connection or lack thereof people have with the natural world. Could local agriculture bridge the gap between the food we consume and where it was produced, resocial-izing people enough for them to know the social origins of the stuff that sustains them? To assist me in this endeavor I propose a synthesis of two concepts that both originate in the field of economics. The first is *human capital* – this refers to our accumulation of knowledge, skills and expertise which we acquire mainly through our education and work experience, and which is of particular interest to potential employers when we are on the job market. The second concept is *natural capital* which comes to us from the sub-field of ecological economics (Costanza et al., 1997). The central thrust here is that were we to include the actual cost of nature's contribution to capitalist production – including those things almost never accounted by actual capi-talist like rainwater, clean air, the carbon cycle, daylight – the higher of price of everything would rein in our tendency to overexploit natural resources. Bracketing for the moment the question of whether capitalist profits could even exist in such a scenario, making transparent the exploitation of natural resources through capitalist production and, hence, our own consumption within this system is, I would assert, the right thing to do.

By synthesizing *natural human capital* from these two concepts originating in economics, I am also attempting to capture the ecofeminist notion of "holding" mentioned in the previous chapter which regards our experiential knowledge of the natural world as a thing of value. This seems especially relevant at this moment in history when so few people are involved in agri-cultural production, and many of us have no idea how the food on our plate even got there.

The Intervale: a local incubator for sustainable agriculture

Vermont, as it turns out, is an excellent place to examine the themes of food equity, social integration, and natural human capital. In 1986, a collective of environmentally minded locals lobbied to have a city dump on the northern part of Burlington transformed into a farming "incubator," a non-profit or-ganization where land and technical knowhow would be made available to train those willing to learn how to do small-scale organic agriculture. By the first decade of the new millennium, 11 independently managed farms were producing over 500,000 pounds of produce annually on the site which is now known as the Intervale Food Hub. In addition to the market-oriented farms, the Intervale was also the site of a CSA collective and a 165-plot community garden managed by Burlington Parks and Recreation. My research plan was thus straightforward: To what degree are the ideals of food equity, social integration, and natural human capital attained in each of these three varieties of local agriculture found at the Intervale – organic market farms, CSAs, and community gardens?

With regard to *food equity* and the people participating in each program, community gardening at the Intervale includes people whose socioeconomic status looked most like that of people in the surrounding area. The relatively low annual fee – $50 at the time – and the attraction for people who do not have space to garden at home assures that community gardeners tend to look more like average citizens, albeit with fewer children. CSA members, in contrast, appear more likely to have children, but are also of a higher socioeconomic status, as reflected in the very high rate of college completion – over 90 percent at the time – among them. The up-front payment of $400 at the beginning of the season (though an installment plan is now available) likely turned many people away. Also, in the case of the Intervale CSA, membership has been promoted largely by word of mouth, assuring, however unintentionally, that a select group of people will actually become members. Organic market farmers have a somewhat more mixed relationship with the question of food equity, their participation in farmers' markets being their saving grace as it allows them to sell their produce at or near wholesale to a more diverse segment of the population. It should be stressed that all three programs at the Intervale – CSAs, community gardening, and organic market farms – are involved in gleaning activities that send unused or excess produce to local social service agencies, including the Chittenden County Emergency Food Shelf.

Comparing the quality of *social integration* among the three different modes of local agriculture at the Intervale, we might argue that organic market farms represent the least socially integrative since they do not bring with them the explicit mission of community involvement that community gardening and CSAs do. That said, the Saturday morning farmers market in downtown Burlington makes clear that organic market farms are not purely about instrumental business transactions. The diverse cross-section of people circulating and chatting among vendors in City Hall Park as local musicians perform on the green represents one of the more hopeful sources of social integration and community-building to be found in the United States today.

A somewhat similar, if less diverse, scene may be found twice a week at the Intervale CSA pick-up point. There, well-educated, mainly middle- to upper-middle-class families, many of whom already know each other, converge in part to pick up their food share, but also to meet old friends and catch up on local goings-on.

Meanwhile, just up the road, a social scene unfolds more gradually. Men and women, some by themselves, some in pairs or small groups, work quietly in sight of each other, pulling weeds, watering, working the soil. Every so often a conversation will break the silence between neighbors, but for the most part people are trying to get gardening done. One might even surmise from this description that community gardening promotes relatively asocial behavior. A *natural human capital* perspective, however, would suggest otherwise. True, a passing observer would notice a lot less conversation going on at the community garden than at the CSA pick-up or the farmers' market. Nonetheless, something essentially social is taking place when people work together, using the knowledge they have gained from experience and from each other, to grow their own food.

This is a central characteristic of natural human capital – it does not occur in a vacuum but, rather, requires an ongoing interaction with both the natural world and with other people. More so than CSAs and organic farms, community gardens provide a context of socialization in which this combination of factors comes together and creates natural human capital. Sharing information about the best methods for organic growing, putting those methods into practice through the physical activity of working in the garden two or three times a week, and taking communal responsibility for the care and use of tools and common areas are but three ways community gardening structures the social acquisition of natural human capital among its participants.

Addressing differential access to sustainable institutions

Though the global market of food production affects us all, class, culture, and race differentiated responses to it are evident. My research and other studies suggest the participants in CSAs and direct-market organic farming tend to be college-educated and of middle-class origin (Hinrichs & Kremer, 2002). In their book, *Cultivating Food Justice: Race, Class, and Sustainability* (2011), sociologists Alison Alkon and Julian Agyeman also argue that dominant voices in the local food movement, especially as established by writers such as Michael Pollan, Eric Schlosser, and Barbara Kingsolver, have tended to be white and highly educated. This is not to say the intentions of these authors are malevolent or wrong-headed. It does suggest, however, that some of the longest standing traditions of small-scale, local food production among African American, immigrant-origin, and indigenous communities have gotten neither the attention nor the policy support they deserve (Hondagneu-Sotelo, 2014; McCutcheon, 2013; Minkoff-Zern Peluso, Sowerwine, & Getz, 2011; Norgaard, 2019).

Food sovereignty is the term Teresa Mares and Alison Alkon (2011) use to describe the move towards local alternatives to the politically influential preferences of industrial agriculture. Originating among farmers in the Global South who have had to compete with highly subsidized industrial farms in the north, food sovereignty activists fight for the rights of local people around the world to determine their own food systems; i.e., sovereign foodways outside the control of domestic and global trade policies that favor large scale production over quality and community. Land reform and moratoria on the patenting of indigenous seed varieties are two examples of policy changes that could help sustain a truly local food movement. Subsidies that favor small-scale production would help, as well.

On the consumer side, an effective strategy for promoting broader integration in local agriculture is through the targeting of lower-income families with government subsidized farmer-to-family coupons and the placement of farmers markets in working-class neighborhoods. These two strategies have met with some success in Burlington and other towns and cities across the country. As opposed to well-meaning charitable donations to the local food shelf, farmer-to-family programs integrate lower-income families directly into the social transactions and exchange of the local market. The literature on social integration and social capital has long argued that a rich civic life and community participation must have at its base the regular opportunity for face-to-face interaction among community members (Mills & Ulmer, 1970; Polani, 1944; Putnam, 2000).

There is reason to believe the local food movement will be a key element in our ability as a species to face uncertain socioecological times. An early indication of this may be found in the examples of local agriculture described above. As in many other social movements, early successes have been dependent on an ongoing process of planning and deliberation among the respective members, partners, and individual producers within each organization. Though an ethical view which values small-scale over industrial agriculture often mobilizes people to become involved in local food production as a matter of principle, it is the institutional forms of local agriculture – CSAs, organic market farms, and community gardens, and the face-to-face relations inherent in them – that form the organizational basis of the movement. In that way, they represent concrete examples of how social movements work to produce and sustain the institutional context necessary for socialization into a more ecologically aware and sustainable future.

There is of course a longer view of the significance of social integration for the success of social movements in the sociological literature. In that vein, we will now turn to the question of what makes people want to work together for a common purpose in the first place. Near the end of this chapter, we will also consider the obstacles elite forms of solidarity place in the way of those who wish to challenge status quo routines deemed unjust or detrimental to our collective future.

Solidarity and the shift from rural to urban life

The world has changed immensely since 1893 when Emile Durkheim, a former rabbinical student turned amasser of urban data sets, first published *The Division of Labor* (2014). In this book, Durkheim argued there was something altogether different and new about the way people worked together and trusted each other in modern society. Whereas in pre-industrial societies collaboration and trust were grounded in lifetimes of interaction and inter-dependence among friends, family, and neighbors – what he called *mechanical solidarity* – the majority of people living in the burgeoning cities of 19th-century Europe had no such social legacy to fall back on.

This was both cause for alarm and cautious optimism, and much of Durkheim's writing lays out evidence-based arguments for both sides of this sociological coin. On the alarming side is his groundbreaking work on the social context of self-destructive behavior, *Suicide* (1951), first published in 1897. Being himself a chief promoter of sociology as a bonafide science, Durkheim's choice of suicide as a topic of research was both resourceful and shrewd. Resourcefully, he exploited some of the underlying fears in modern life, especially among leaders in regional and national governments who were by the end of the 19th century collecting substantial amounts of information on the behavior of their citizenry. Of particular interest was crime, causes of death, and the many vices that were thought to undermine the proper comportment of a civilized society.

Durkheim did not so much share the fear-based rationales of government bureaucrats as he recognized that for the first time in human history a systematic accounting of human behavior could be measured, even tested, in light of underlying theories about what motivates groups of people to do the things they do. These data points indicating a person's age, religious identity, criminal record, and marital status, among other things, are what Durkheim referred to as *social facts* – i.e., the empirical building blocks of the emerging academic discipline of sociology. Shrewdly, he deployed these social facts to bolster a straightforward sociological argument: Suicide, a quintessentially individual act, could be explained on the basis of social context. Namely, different levels and kinds of social disintegration are good predictors that individuals from certain socially defined groups will have a greater likelihood of taking their own life than those from other groups.

Anomie is the term Durkheim used to describe this lack of connection to others; a not unexpected outcome in the modern setting where, in contrast to pre-industrial times, the work we do is highly specialized and our relationships with others, especially in the economic sphere, can be fleeting. In this new era, *organic solidarity* is what holds together the complex, dynamic, and rapidly changing social context of the modern age. More often than not, we now find ourselves dependent for our survival on technology, shelter, and food produced for us by people we probably will never know. Remarkably, our unspoken trust in anonymous engineers, far away farmworkers, computer programmers, water treatment plant operators, and millions of others with specialized knowledge and skills we ourselves will likely never attain is what permits us to lead the lives we live – a miracle of faith, you might say.

This edifice of complexity and interdependence we all inhabit, however, comes at a social cost. To wit, the built-in ties of community, social obligation, religion, and other homogeneous social traits found under mechanical solidarity are now few and far between. For Durkheim, it is the mismatch between, on the one side, the human need for continuity and moral guidance found in stable communal ties and, on the other, the often transient, market-friendly ties of commerce and finance that leads to a personal sense of anomie, i.e., a lack of meaningful integration with others. At the societal level, this meant that the *collective conscience* of common norms, beliefs, and values becomes frayed.

Despite this somewhat dreary assessment of life under industrial capitalism, Durkheim was hopeful that modern people living within the milieu of organic solidarity would find a way to re-integrate and establish a meaningful basis for common social norms and trust in a complex society. To be clear, he did not believe people in the contemporary setting would ever return to an idealized earlier time of mechanical solidarity. Instead, Durkheim felt that *voluntary associations* grounded in occupations would take on the role once filled by local institutions of religion, family and municipal government.

Though there were no guarantees, social integration might once again be attained through the deliberate creation and maintenance of voluntary associations that, in addition to giving people a modicum of moral guidance and sense of common purpose, would also work against the undemocratic tendencies of representative democracy which Durkheim saw as a looming threat. When people met and spoke with each other on a regular basis about common concerns and grievances, democracy could be grounded in civic participation. In providing this function, voluntary organizations served as a necessary bulwark against the creeping bureaucratic and despotic potential of socially distant representative governments.

Durkheim's empirically grounded research on suicide established him as one of the leading visionaries behind the establishment of sociology. However, it is arguably his theoretical work on the role of associations in structuring social integration that has had a more lasting influence on researchers seeking to understand the animating source of group solidarity, especially during times of dramatic social change. This could include everything from how do CSAs and community gardens promote social integration among neighbors, to how can groups of average working people possibly hope to stand up against the repressive forces of a dominant class?

Sources of solidarity in the French Commune of 1871

In this last regard, one of the more fascinating instances of this occurred in Durkheim's own country of France. In the winter of 1871, 82 years after the French Revolution, urban denizens of diverse social class backgrounds found themselves running the city of Paris. This unprecedented four-month-long experiment in local municipal governance became known as the Paris Commune. After the fall of Napoleon III and the Second French Empire in 1870, the conservative French Army attempted to take back weapons from artillery parks defended by the neighborhood-based National Guard. The guardsmen, who had no interest in welcoming the monarchy back to power, held strong and forced the army to retreat to Versailles before returning in late-May to crush the communalist insurgents.

Over a hundred years later, Roger Gould, a sociology graduate student at Harvard, was becoming frustrated with sociological theories used to understand how and why people become committed to political movements for social change. The dominant approach, he felt, lacked empiricism and relied too heavily on Marxist thought which saw political commitment as a direct function of social class status. The Paris Commune, it was thought, was a kind of historical replay of the French revolution, only this time the working class had taken up the adversarial role played by the bourgeoisie against the aristocracy in 1789. Influenced by James S. Coleman's social network theories and Theda Skocpol's historical sociology work on revolutions, Gould wanted to show that armed resistance and, hence, social movements more broadly, cannot be based on socioeconomic class, alone. Archival data from 1871 of Parisian neighborhoods provided the ideal basis for demonstrating the pivotal role multiple social ties play in generating commitment to a social movement.

On the face of it, social organization during the Paris Commune was simply a matter of urban geography with place-specific identities and loyalties tied to the residential zone or *arrondissement* in the city where people lived. Upon closer inspection, however, Gould found salient ties within arrondissements to the neighborhood-specific National Guard battalions in which residents served during the previous two decades of the Second Empire. It was these overlapping ties of neighborhoods and military service – sometimes coinciding, sometimes linked across arrondissements – that best predicted commitment to armed resistance against the French Army (Gould, 1991).

Commitment, in this case was measured by the death rate in each arrondissement due to military repression. Parisian residents were much more likely to resist to the death the monarchist-led army if they came from arrondissements with large National Guard battalions. This was especially true if they came from neighborhoods with many crossover linkages through arrondissement-specific battalions containing members who resided in other arrondissements. These overlapping networks of social ties wherein one can know the same person in different social capacities (neighbor, soldier, coworker, for example) are what sociologists refer to as multiplexity.

Multiplex ties in the American Civil Rights Movement

The Paris Commune underscores the ways in which social integration, structured by membership in previously existing associations can serve as a powerful interactive basis for trust and solidarity among people in a modern setting. Multiplexity, really a more contemporary conceptualization of Durkheim's social integration, can also help us understand more recent collective action triumphs, as well. Though he did not use the network language of multiplex ties, Aldon Morris' research on the *indigenous organizational resources* of Black communities in the South provides an illuminative example of how overlapping connections to the same people within local institutions sustained the early successes of the Civil Rights Movement.

Theoretically, Morris' indigenous perspective provides a critique of two dominant approaches in the literature on social movements (Morris, 1986). Sociologists writing from a *collective behavior perspective* tended to discount the degree to which planned, rational organization was a determining factor in the emergence of protests and defiant political actions on the part of subordinate groups in society. Emotional and irrational psychological responses to social strain, it was argued, were what drove people to engage in "mob" behavior, and other defiant activities outside the approved comportment of the dominant culture.

Coming from a different angle, *resource mobilization* theorists were beginning to challenge collective behavioralists in the 1970s with research that highlighted funding and the outside support required of social movements if they were to be successful beyond the very local level. From this perspective, the primary goal of community-based movements was not to call attention to unjust treatment on the part of the powerful, but rather to gain the sympathy and material support of outside elites. This was necessary since, by definition, the powerless lack political organization. Morris' account of Black community political organization during the first decade of the Civil Rights Movement found both collective behavioralist and resource mobilization perspectives wanting.

Somewhat ironically, the history of forced segregation and violence against Blacks in the United States had, by the mid-20th century, fostered a high level of internal organization, reflected most obviously in the role of black churches, their congregations, and ministers who, as Morris points out, did not receive their paycheck from the dominant white population. Severe segregation meant black communities often had within them a large degree of socioeconomic diversity such that menial laborers, schoolteachers and doctors might all be part of the same congregation. Common grievances, lives segregated from whites, and tight-knit communal bonds were the fertile ground upon which an insurgent movement could grow and attain political vitality.

Thus, before outside financial support was available, bus boycotts – notably in Baton Rouge, Louisiana and Montgomery, Alabama – could be carried out and communicated efficaciously, beginning in the pulpit on Sunday morning, and spreading out through neighborhood-level networks where protests and carpools for work could be coordinated directly. Emotions, common culture, and religious faith were clearly part of what sustained the strength of this movement through its legislative successes of the 1960s. However, especially in the context of high-risk activism where challenges to Jim Crow-era segregation were met with violence and murder, multiplex connections of trust and interdependence were indispensable.

Coming from a different angle, sociologist Doug McAdam examined the determinants and long-term effects of high-risk activism among middle-class white college students during what came to be known as Freedom Summer

(McAdam, 1986). As the strategic use of non-violent protests in the South garnered national attention and sympathy from outside, many college-aged students from around the country felt the need to do something to support the movement against racial injustice. An opportunity to contribute arose in the summer of 1964 when Civil Rights leaders organized an ambitious effort to register blacks to vote in Mississippi, a state that had long suppressed black political participation through poll taxes, literacy tests, and the threat of violence.

Though many young people from outside the South supported the struggle for civil rights and wanted to help, not many could or did. Using a unique set of data – volunteer applications submitted to assist in Freedom Summer voter registration – McAdam wanted to know what, in the case of white middle-class college students, were the determining factors for engaging in high-risk activism. More specifically, among those who had submitted applications, who actually showed up to volunteer in Mississippi and who did not? Those who did volunteer, he found, shared a number of common characteristics. First, they were "biographically available"; i.e., they tended to be young, single, and had enough financial means to both make the trip away from home and not have to work over the summer. Second, though presumably all the applicants believed strongly in the ideals of equality and social justice espoused by the Civil Rights Movement, what really distinguished those who showed up in 1964 was their network ties. That is, they were both much more likely to have previous existing ties to Civil Rights organizations, and to know someone else who was also participating in Freedom Summer activism than the no-shows.

It should be noted that these single-layer ties to movement organizations and fellow activists are quite different from the multiplex ties of community described by Morris, based on histories of interactions among people occupying multiple roles within local institutions. That said, a follow-up set of interviews to the original study found that, 25 years later, Freedom Summer activists had continued living lives dedicated to activism and fighting social injustice, often tapping into institutional and non-profit network ties established during the Civil Rights era. Thus, what began as a youthful desire to bring about positive change in the world resulted for many in a life-changing experience, influencing people's career choices and shaping longtime networks of trust among friends and fellow activists.

Social closure and positive educational outcomes

During the same year as Freedom Summer, sociologist James S. Coleman was tasked by the Johnson Administration with the design and execution of the most extensive survey ever taken of US students, their families, and teachers. Commissioned a decade after the *Brown vs. Board of Education* Supreme Court decision declared public school segregation unconstitutional, the goal of the 'Coleman Report' was to understand in a comprehensive fashion the underlying causes of persistent unequal outcomes in the American public school system (Marsden, 2005). The results of this study were both insightful and contentious. At a time when activists were pointing to the unequal distribution of opportunity inherent among racially segregated schools, Coleman's research team argued that resource differences between schools were a relatively insignificant predictor of student academic outcomes when compared to the disparate resources families brought to bear in the household. From that perspective, relevant questions included, was a child being raised by two parents? Did students have access to books and other educational materials at home? And, what was the highest level of education attained by the head of household?

At the time, some critics accused Coleman of blaming the victim for persistent racial achievement gaps in the schools and providing fodder for "separate but equal" segregationists. He was, in fact, very much in favor of school integration – just not for the reasons Civil Rights activists expected. African American and other socioeconomically disadvantaged students in the public school system, he argued, benefitted socially and academically from being around fellow students from families of greater advantage than their own. The diversity of interaction across categories of race and social class – not access to higher quality school facilities – was the crucial benefit integrated schools offered disadvantaged students, building common conduits of academic culture and practice that had been deliberately cut off through forced segregation.

Coleman would go on to develop further his "social capital" theory of educational attainment, rendering a more nuanced account of how overlapping ties of family, institutions, and community bolster student academic performance. Using data from the 'Beyond High School' survey taken in the early 1980s, Coleman found that students in Catholic and other-religion affiliated high schools were much less likely to drop out of school than students in either public or non-religious private schools (Coleman, 1988). At the household level, he also found that, controlling for parents' education and financial capital, students from families with a single parent or from families with many children were more likely to drop out than families with two parents or with a single child.

The underlying theme connecting these predictors of academic success or failure is *social closure* which, regarding high school teens, really comes down to adult supervision. Essentially, when it is almost certain that no one parent can ever know at all times what their teenage children are up to, it helps tremendously to be in a community where adults know each other and have established some degree of interpersonal trust and communication. This is most likely to happen when, according to Coleman, not only do I know and trust you through regular interaction, but I know other people who know you in this way, as well – a kind of social triangulation. In the household, this explains the intrinsic child-monitoring advantage a two-parent family has over a single-parent family.

Outside the home, institutional ties can also facilitate forms of social closure. This, according to Coleman, is primarily why Catholic and other religion-based high schools have lower dropout rates: The additional layer of institutional affiliation through religious services, volunteer committees, and social events provides greater opportunity for parents to know the constellation of people around their children in a variety of ways. For the sake of contrast, imagine a scenario where the people parents know at work, in their religious community, and through their children's school constitute a wide web of people without overlapping ties. Though these parents may have diverse extended social networks, they likely have little in the way of social closure with respect to the supervision of their children.

The cultural and emotional glue of multiplex ties

In sketching the basic geometry of social closure through his work on education, Coleman provides a theoretical basis for conceiving of and understanding multiplex communities. More recently, sociologists have fleshed out these network models by drawing attention to the emotional and identity content of social ties in a multiplex context. That is, what is the glue that bonds people together in communities of trust and interdependence? In this regard, close tie relationships have proven to be especially adhesive. For example, if a person has a strong friendship with two different people, it is highly unlikely that these two friends will not themselves be on friendly terms with each other. In fact, a violation of this principle has been referred to by Stanford sociologist Mark Granovetter as the "forbidden triad" – you cannot be socially unconnected to your good friend's other close relationships (Granovetter, 1977). Put another way, close ties beget more ties which, as they accumulate, become the basis of multiplex communities. But what makes these close ties stick?

One angle on this has been to focus on culture. Sociologist Omar Lizardo's research suggests that the strength of one's social ties is influenced by one's preference for popular versus "highbrow" culture (Lizardo, 2006). According to his work, the lived environment of popular culture based on sports, entertainment, and the consumption of mainstream products serves as a bridge bringing together people across categories of difference. Weak tie connections and social interactions are thus disproportionately based on reference to popular culture, things that everyone can recognize and comprehend. By contrast, highbrow culture – things that by definition require commitment, time, and education dedicated to the understanding of esoteric themes – is more directly tied to strong tie relationships. That is, a common understanding of artistic, literary, and/or intellectual topics typically inaccessible to non-elite social groups can bring together people with more refined tastes in close-tie bonds.

Though Lizardo focuses on highbrow culture as a basis for close-tie bonding, this applies primarily to societies saturated in consumer culture where only a privileged and highly educated minority can afford to step outside the popular mainstream. As was made evident by the Paris Commune and Civil Rights actions described above, however, non-elite alternatives to building close-tie, multiplex communities are clearly feasible, as well. What those two social movements underscore is the significance of local knowledge of social networks, regional geography, religious affiliations, and neighborhood histories. To give another pertinent example, during the Prague Spring of 1968, Czech resisters removed or obscured road signs as an effective way of confusing occupying Soviet troops. As anyone who has spent time in a foreign country knows, the knowledge of what remains of local cultures can be much more confounding for outsiders than elite knowledge of, say, a Verdi opera. The original close-tie culture is local culture and it is only in a context where mass popular culture has to a large degree displaced local culture that highbrow culture appears as a fruitful path for some to close-tie networks.

Today, the difference between elites brought together on the basis of a preference for aged cheese and vintage wines versus, for example, families living in a multi-generation Mexican American neighborhood where people likely attend the same schools, are in the same religious congregation, and possibly work for the same employer is multiplexity. In the Mexican American enclave, people do not come together because of their refined preferences for high-end consumer goods, but because they rely on each other for mutual support and share common cultural traits that facilitate trust and interdependence. Moreover, in contrast to the status affirmation bestowed upon folks who can afford to smoke expensive cigars and drink single-malt scotch, social closure is the principal benefit to membership in a multiplex community. That is, status is secondary to shared histories and solidarity in the non-elite multiplex context.

Undermining the ecology through integrated communal ties

Of course, the mere presence of multiplex communities does not mean that the people within them will always act in ways that benefit the greater good, or even their own self-interest. In her book, *Strangers in Their Own Land*, sociologist Arlie Hochschild has written about communities in Louisiana with deep historical roots, a common ethnic culture, and active participation in civic and religious organizations (Hochschild, 2018). Their deep sense of place-based solidarity unfortunately blinds them to both the deleterious health and environmental impacts the petrochemical industry has had in their communities. Moreover, the shared belief that far away government officials have long neglected their concerns collectively dismisses the possibility that government regulation may actually improve their circumstances. In this worldview, only industry, incentivized by state government and relatively unmonitored by the federal government, can save the regional economy; an economy hit particularly hard by layoffs in the public sector where tax-based incentives for industry have resulted in drastically reduced government revenue. In this case, multiplex insularity – a lack of weak-tie links to contrasting points of view – may be part of the problem.

Also at issue in Hochschild's study is the role of elites in reinforcing this insularity. Along with siting polluting industries near populations deemed "least resistant" – e.g., longtime small-town residents, high school-educated only, no prior history of activism, politically conservative – Hochschild provides evidence through interviews and the public record that the petrochemical industry hired undercover agents to infiltrate and stymie attempts at environmental activism among concerned citizens in her study. Beyond these more targeted strategies, Hochschild reminds us that conservative elites – her ethnographic work highlights Donald Trump's 2016 presidential campaign as an exemplar, in this regard – have effectively tapped into the "deep story" of injustice perceived by many working-class, conservative, white Americans. This highly divisive portrayal of the national context goes something like this: there are those in society who have unfairly "cut in line" to attain the American Dream, and the federal government – through social welfare programs, hiring preferences, and lax immigration laws – has facilitated these unjust maneuvers which disproportionately benefit non-whites and women. Ideologically, non-elites across race, ethnic, and gender backgrounds are thus divided in ways that make political challenges to industrial practices threatening both the natural environment and public health quite difficult.

Hochschild's recent work contributes to a long tradition in sociology of drawing attention to the role elites play in shaping public opinion and undermining challenges to the status quo. Karl Marx's notion of "ideology," for example, describes the way religion, nationalism, and other elite-imposed worldviews distract people from their common interests as workers in rising up against exploitation under the capitalist system (Marx & Engels, 1965). As mentioned in the previous chapter, this same idea would inspire the Frankfurt School to understand first the persuasive role of propaganda in promoting anti-Semitism under the Third Reich in Nazi Germany, and later the growth of the "culture industry" in the United States to promote consumer capitalism (Adorno, 2005). Along those lines, current research in environmental sociology has sought to bring to light the efforts corporations have made to skew the debate about climate change.

In a 2016 paper published in the *Proceeding of the National Academy of Sciences*, Justin Farrell uses social network analysis to assess the impact of corporate funding on both the content and quantity of texts produced by 164 organizations involved in the climate change countermovement from 1993 to 2013. Relative to those without corporate support, corporate-funded organizations were much more likely to produce polarizing content, arguing, for example, that carbon dioxide is beneficial to the environment, that temperature fluctuations reflect long-term natural cycles, or that the economic costs of climate change policy far outweigh any ecological benefit.

Beyond his analysis of countermovement texts, Farrell also includes controls for *corporate interlock* – the overlapping ways members of corporate boards simultaneously hold directorships on multiple boards. This elite utilization of multiplex ties greatly facilitates both the formulation of a common critical discourse on climate change and its dissemination through think tanks and other industry-sponsored foundations to the general public. According to Farrell, ExxonMobil and the Koch family foundations have been major backers of this strategy.

At this point, it would be tempting to slip into an easy cynicism about how corporate leaders through their network-based collusion can instrumentally manipulate the public into going along with whatever serves their needs for generating greater profits, even at the cost of public health and global biodiversity. However, returning to Hochschild's central argument in *Strangers in their own Land*, by attributing political dissonance in American life entirely to elite manipulation, we miss the lived experience of average people which arguably more directly informs their position on issues such as environmental protection and affirmative action. It is not that we have to agree with those whose opinions we find inaccurate if not offensive, but we will not make progress towards understanding if we simply believe they have been bamboozled and we have not.

As evidenced by both the elite and underdog successes highlighted in this chapter, multiplex ties appear to have something to do with efficaciously bringing about desired changes in society. It also may appear through the Farrell study and others like it that elites always have the upper hand when it comes to multiplex ties. The problem for elites, however, is that, by definition, there can only be so many elite individuals and, on the whole, their lived experience is not that of the average person. To the extent non-elites actually interact with each other, we can presume they will relate to and understand each other more so than they would elites.

True enough, charismatic elites may occasionally be able to tap into a common sense of injustice and exploit it for ends that do not benefit the greater good. That, however, is an outcome most likely in society where people do not trust each other more than they would a distant leader; or, perhaps just as bad, only trust a limited number of people who happen to share a common set of sociocultural characteristics. This particularized form of trust appears to present a looming threat to many democracies around the world wherein the sense that a dominant group with a culturally defined set of acceptable characteristics can be trusted while other groups cannot. In these cases, generalized trust has been jettisoned in favor of scapegoat politics. History tells us this approach cannot end well. Sociology suggests this is not destiny and, in fact, there is much we can do to attain a more favorable outcome. As we shall see in the next chapter, along with nurturing our own multiplex ties to others, we also need to be on guard against corporate incursions that divide us from each other and our interdependent relationship with the natural world.

Glossary

Anomie   Emile Durkheim's term for the experience of lacking social connection and meaning in life due to a lack of social integration with others.

Biographic Availability   An intersection of characteristics including age and socioeconomic status that make certain individuals more likely to engage in political activism than others.

Collective Behavior Perspective   The theory that political protests and demonstrations are driven more by situational group impulses and reaction than planned, rational organization.

Collective Conscience   The Durkheimian idea referring to a shared societal understanding of how the world works based on a common system of norms, beliefs, and values.

Ecological Feminism   A theoretical perspective which critiques male-dominated industrial capitalism, and valorizes "moral economies" of the household and the "holding" of cultural traditions which emphasize continuity, preservation, and the conservation of natural resources.

Food Equity   The idea that all people should have access to plentiful and nutritional foods, and that those goods consumed should be the product of fair and ethical food production.

Food Justice   The social movement founded on the idea that all people of different socioeconomic backgrounds should have equitable access to a variety of nutritious food.

Food Sovereignty   The social movement and pattern of choosing local food alternatives to industrial agriculture.

Human Capital   One's collection of knowledge, skills, and expertise, mainly developed by experience in the workplace or in educational settings.

Mechanical Solidarity   The distinct social relationship individuals have with others who live similar lives and hold similar beliefs and have similar skill sets relative to their own; indicative traditional rural and agricultural societies.

Multiplexity   The quality of having overlapping social ties; knowing someone in more than one role or social category.

Natural Human Capital   The social benefit that results from building healthy relationships both with the environment through interactions with other humans.

Network Ties   In social network theory, the basic units of connection of one person to another.

Organic Solidarity   An element of Durkheim's social theory referring to a human tendency to rely on people with specializations different than our own; indicative of our interdependence with others in modern life.

Resource Mobilization Perspective   Social theory which argues successful attempts at collective action to promote political change require organization and structure to be successful, especially as relates to the ability to garner support from sympathetic outside groups with resources at their disposal.

Social Closure   A social triangulation of knowing individuals through different network ties, thus bridging gaps of social support and supervision in the community.

Social Facts   For Durkheim, these referred to measurable aspects of social life that provide the empirical basis for a science of society.

Voluntary Associations   Organizations in modern life outside workplace or household based on common interests that, according to Durkheim, are essential for sustaining a modicum of social integration in the context of organic solidarity.

Questions

1   According to Durkheim, one of the best ways to stave off the looming threat of anomie in modern life and thus promote organic solidarity is to join a group or organization. In which kind of groups or organizations have you been or are you currently a member? Think of a cause or issue you are currently interested in. Now get online and find an organization nearby in your university or community involved in this topic. Would you consider joining? Why or why not?

2   Think of three people in your life who are not family but to whom you are close. Now draw a small diagram with dots identifying the four of you, and lines indicating strong relationships between the dots where they exist. Do your friends know each other? Could this network cluster be described as having social closure? Also, is this a multiplex cluster? That is, do people within this network know each other in a variety of different context (as students, coworkers, club members, etc.)?

3   Sociologist Omar Lizardo has argued that "high brow" culture facilitates the formation of close ties by its very nature since only a select few in the population have the time and resources to dedicate to the cultivation of elite tastes. What do you think? What is the cultural basis of your closest relationships? Can you think of examples wherein "low culture" experiences foster close tie relationships?

References

Adorno, T. W. (2005). *The culture industry: Selected essays on mass culture*: Routledge.

Alkon, A. H., & Agyeman, J. (2011). *Cultivating food justice: Race, class, and sustainability*: MIT press.

Coleman, J. S. (1988). Social capital in the creation of human capital. *American Journal of Sociology, 94*, S95–S120.

Costanza, R., d'Arge, R., De Groot, R., Farber, S., Grasso, M., Hannon, B., . . . Paruelo, J. (1997). The value of the world's ecosystem services and natural capital. *Nature, 387*(6630), 253.

Durkheim, E. (1951). Suicide: a study in sociology [1897]. *Translated by JA Spaulding and G. Simpson (Glencoe, Illinois)*: The Free Press, 1951.

Durkheim, E. (2014). *The division of labor in society*: Simon and Schuster.

Farrell, J. (2016). Corporate funding and ideological polarization about climate change. *Proceedings of the National Academy of Sciences, 113*(1), 92–97.

Gould, R. V. (1991). Multiple networks and mobilization in the Paris Commune, 1871. *American Sociological Review, 56*(6), 716–729. doi:10.2307/2096251

Granovetter, M. S. (1973). The strength of weak ties. *American Journal of Sociology, 78*(6), 1360–1380.

Hinrichs, C., & Kremer, K. S. (2002). Social inclusion in a Midwest local food system project. *Journal of Poverty, 6*(1), 65–90.

Hochschild, A. R. (2018). *Strangers in their own land: Anger and mourning on the American right*: The New Press.

Hondagneu-Sotelo, P. (2014). *Paradise transplanted: Migration and the making of California gardens*: University of California Press.

Lizardo, O. (2006). How cultural tastes shape personal networks. *American Sociological Review, 71*(5), 778–807.

Macias, T. (2008). Working toward a just, equitable, and local food system: The social impact of Community-Based agriculture. *Social Science Quarterly, 89*(5), 1086–1101.

Mares, T. M., & Alkon, A. H. (2011). Mapping the food movement: Addressing inequality and neoliberalism. *Environment and Society, 2*(1), 68–86.

Marsden, P. V. (2005). The sociology of James S. Coleman. *Annual Review of Sociology, 31*, 1–24.

Marx, K., & Engels, F. (1965). *The German ideology*: International Publishers, New York.

McAdam, D. (1986). Recruitment to high-risk activism: The case of freedom summer. *American Journal of Sociology, 92*(1), 64–90.

McCutcheon, P. (2013). Community food security "for us, by us": The Nation of Islam and the Pan African Orthodox Christian Church. *Food and Culture: A Reader* (pp. 572–586): MIT Press.

Mills, C. W., & Ulmer, M. (1970). Small business and civic welfare. In M. Aiken & P. Mott (Eds.), *The Structure of Community Power*: Random House.

Minkoff-Zern, L.-A., Peluso, N., Sowerwine, J., & Getz, C. (2011). Race and regulation Asian immigrants in California agriculture. In A. H. Alkon & J. Agyeman (Eds.), *Cultivating food justice* (pp. 65–86): The MIT Press.

Morris, A. D. (1986). *The origins of the civil rights movement*: Simon and Schuster.

Norgaard, K. M. (2019). *Salmon and acorns feed our people: Colonialism, nature, and social action*: Rutgers University Press.

Polani, K. (1944). *The great transformation: The political and economic origins of our time*: Rinehart.

Putnam, R. D. (2000). *Bowling alone: The collapse and revival of American community*: Simon and Schuster.

5 Fetish, rifts, and farce

Alienation from what we make, buy, and toss

One of the principal hurdles in the way of a more sustainable future is the dominant culture of consumption we have all, to one degree or another, been socialized into. In this chapter, we will explore in greater detail how the things we consume can distort our perception of material and social reality. The term Karl Marx had for this misperception was *commodity fetishism*, a form of social alienation which, frankly, is much less kinky than it sounds. If your mind is a field of understanding through which you perceive and try to makes sense of all that surrounds you, consumer products are like large objects with their own gravitational pull, capable of drawing you away from perhaps less immediately attractive but no less consequential elements in the world around us.

As we learned in previous chapters, advertisers and marketing firms have long understood the significance of childhood socialization and, more broadly, the way media context can shape our attitudes and desires. A century before legions of creative economy workers were harnessed to persuade the masses to buy, however, Marx argued that the things themselves produced through industrial capitalism – commodities – distract us from the human effort that went into making them. We fixate, as it were, on the object, which of course makes perfect sense. If I am in the market for a new smartphone, I consider above all the functionality of the product. How big is the screen? What quality photos can it take? How long will the battery last? Will its processing power be sufficient for gaming and watching HD videos? And, perhaps most importantly, how much will it cost? At the very least, savvy consumers will want to have a working list in their head about basic expectations for a costly product of this sort.

The problem, according to Marx, is that by "fetishizing" the product – i.e., focusing only on its tangible attributes – we ignore the social origins of its production. What, for example, are the working conditions of the people laboring in the factories that manufacture these things? How much are they being paid relative to the profits earned by the manufacturer? Relatedly, why are these products that the vast majority of American adults possess not being produced in the United States? At the end of the smartphone's lifespan (about three years, according to the manufacturers) where do its remains – also

DOI: 10.4324/9781003110668-5

known as "e-waste" – end up? Which communities are disproportionally impacted by the processing of used electronic products for "recycling"; itself an often highly toxic endeavor?

This last line of questioning has major implications for social inequality and environmental justice at the global level, but is rarely if ever brought to our attention at the moment of purchase. This is in part by design – why on earth would anyone trying to close a deal on the sale of an automobile, for example, want to bring up serious concerns about greenhouse gas emissions and unfair labor practices in another part of the world? Commodity fetishism, however, also points to our own complicity in the consumer transaction – we ourselves do not want to be reminded of these inconvenient truths at the moment our desires are about to be fulfilled. A fetish, after all, implies gratification will be met through our consummation with the desired object itself (maybe this is kinkier than I thought). To be reminded that consumer products originate in the material world of wages, resource extraction, and worker grievances can really ruin the mood. It seems pretty clear; it is precisely this disconnect between consumer goods – hamburgers, drones, decaf coffee capsules, what have you – and their social origins that keeps the wheels of commerce turning while also permitting many of us to sleep at night.

Conspicuous waste in a rationalized world

But what, exactly, is this consumer gratification people so desperately seek? What is motivating large portions of the population to buy more and more things they do not actually need? (This last point, incidentally, is easily substantiated by the mountains of garbage – about 1,600 pounds per capita annually in the United States, including uneaten food – that end up in landfills (Gunders & Bloom, 2017).) As it happens, quite a few social thinkers have tried to wrap their brain around this question over the last century or so, some more successfully than others. One of the standouts in this regard is the iconoclast American economist/sociologist Thorstein Veblen. At the end of the 19th century, the United States was still another 50 years away from becoming a global economic superpower. Most Americans lived in rural areas, and agriculture was still at the core of the national economy. Veblen's graduate work at Yale at the time was highly critical of industrial capitalism and was met with indifference if not disapproval by his academic superiors. He thus spent much of his early career on his parents' Wisconsin farm immersed in books and economic theory before making his way back to academia as a fellow at the University of Chicago. It is there he completed his most famous work, *Theory of the Leisure Class: An Economic Study of Institutions*, a scathing critique of upper-middle-class consumer culture (2017).

In many ways, *Theory of the Leisure Class* described a world resulting from yet in total contrast to Max Weber's theory of rationalization. As you might recall from Chapter 2, rationalization referred to the drive within industrial capitalism to ever greater levels of efficiency, calculation, and predictability, all in the name of profit maximization. By contrast, the sons and daughters and spouses of successful industrialists – the "leisure class" – engaged in the most ostentatious forms of inefficiency and waste, what Veblen referred to as *conspicuous consumption*. Thus, for the masses there were long hours of monotonous labor to increase industrial productivity. For the elite, however, there was highbrow literature, art collections, horseback-riding, sailing, and other unproductive endeavors that served primarily to convey to others superior levels of social status.

In addition to these overt signals of status attainment displayed through conspicuous consumption, Veblen described other slightly more subtle strategies. *Conspicuous leisure* implies, for example, that although your friends may not be with you as you take your ski vacation in the Swiss Alps, or relax on a Caribbean beach, you can certainly talk about it when you get back. Of course, these days you can actually share directly online your exotic vacations as they happen to the barely suppressed envy of your family and peers back home. You may, in fact, not even be relaxing – a cross-country cycling trip or climb up the Himalayas is, of course, hard work. The important thing is that it be unproductive labor; you are doing it because you want to and you can afford to, not because a capitalist is paying you to do it.

In other words, conspicuous leisure is really about your distance from toil which, as it turns out, is subject to cultural interpretation. There are still some places on earth, for example, where skin darkened by the sun's rays is adamantly avoided. Beyond concern for one's dermatological health, which is certainly valid, and racial undertones, which are certainly possible, many cultures now and at earlier times in our history have associated tanned skin with hard manual labor. Not to be tanned, by contrast, was a clear signal of not having to toil. These days, as far fewer people in economically dominant countries like the United States work outside, a tan may just as well signal your proclivity for unproductive leisure activities.

One final angle on motivations for consumer excess is what Veblen refers to as *vicarious consumption*; that is, the ability of others to use your considerable economic power to spend as they will. In late 19th-century America, Veblen was thinking mainly of the wives of wealthy businessmen who, for one, did not have many options at the time in terms of their own economic independence, but also served by extension through their tastes in clothing, interior decorating, and even regular attendance at church, as an overt gauge of their husband's social status. It is important to keep in mind that Veblen's critique was aimed at a relatively small and elite sector of American life at a time when the "middle class" as we currently think of it did not yet exist. It is only later – in the post-World War II era, really – that Veblen's writing appears to have broader application to other segments of American society. After a period of unmatched economic growth during the middle of the 20th century, it suddenly seemed that vicarious consumption – be it access to your parents' credit card, a summer vacation in Europe, or a family subsidized college education – was a much more attainable or at least aspired way of life.

Cultural capital, or the sociological significance of moldy cheese

Can it really be, though, that the driving engine of consumer capitalism is our fragile egos and desperate attempts to convey to others whom we don't even know our admirable social status? Well, yes. Or, at least that is what contemporary researchers influenced by Veblen's work tend to agree on. A less cynical if highly influential approach is that of French sociologist Pierre Bourdieu. As opposed to simply signaling their social status, Bourdieu believed consumers obtain genuine satisfaction from the things they consumed; that people from privileged backgrounds, for example, actually enjoy going to the opera, reading highbrow literature, and the rarified culinary nuances of moldy cheese and vintage wine. In contrast to more relativistic accounts of consumer tastes, Bourdieu argued that there are real, objective differences between good art and bad art, fine dining and *la malbouffe* (junk food), and in general between "highbrow" and "lowbrow" culture. The determining factor for being able to discern high quality from low quality experience – what Bourdieu calls "cultural capital" – is education (Bourdieu, 2011).

In many ways, there is no real contradiction between Veblen and Bourdieu's theories on consumer culture – consumer tastes can be both a result of years of socialization within a certain cultural terrain, and a way to signal superior social status. In fact, status superiority itself results from the fact that not all cultural and consumer opportunities are distributed equally in a class-stratified society. The fact that most people do not have ready access to a horse makes dressage distinctive from other competitive sports. That said, there is a key point of contrast between these two thinkers. While Veblen appeared motivated by a desire to indicate the waste inherent in elite consumer practices, Bourdieu – especially in his classic tome on Parisian buying habits, *Distinction* (1984) – wants to remind us that social class differences are often viscerally perceived by our contrasting tastes and distastes in art, music, food, and other class-stratified elements of human culture. Does the thought of eating buttered snails nauseate you or make your mouth water? When was the last time you ate white bread? Do you know why some upper-middle-class people in the United States go by the nickname "Trip"? According to Bourdieu, your answer to these and other similar questions may tell you something about your social class status in American culture.

Striving for distinction in a stratified milieu

More recently, sociologist Juliet Schor, has brought together the themes of material waste and social inequality in her research on American consumer culture where status signaling is increasingly financed by debt. As she sees it, the move away from regular interactions with neighbors and other social equals in modern life has only strengthened the aspirational frame of reference through which consumer lifestyle expectations are aligned with the representations of the upper-middle class provided in advertising and the popular media. In her book, *The Overspent American: Why We Want What We Don't Need*, Schor (1999) describes the lengths to which households go to keep up appearances. As middle-income wages began to stagnate in the 1970s, American families tried a number of strategies to maintain what was deemed a respectful lifestyle. As mentioned in Chapter 3, one of the long-lasting changes that took place at this point was the mass entry of middle-class women into the workforce. In this particular shift, there was obviously an aspect of economic liberation for women at the time, but a key determining factor here was the fact that middle-class purchasing power was no longer attainable on a single income. The additional flow of revenue from women's earnings was, for many American families, a necessity.

Another strategy underscored in *The Overspent American*, is the increased amount of time people spend at work, also to keep up with the higher costs of living. During the final three decades of the 20th century, according to Schor, Americans increased their paid hours of employment on average by 163 hours a year – about a month's worth of work. Sociologist Arlie Hochschild gives an account of the emotional toll this shift has had in her book, *The Time Bind* (1997). There she describes the stresses of middle-class family life where both parents work, but women are still disproportionately responsible for raising children and household chores. Not surprisingly, many of the women interviewed for that study find life at work – a haven of recognition and comradery among their peers – much more gratifying than life at home. Hochschild also sees in her ethnography attempts on the part of parents to compensate for the lack of "quality time" spent with children through the purchase of gifts and other consumer indicators of parental love.

A third strategy mentioned by Schor is the greater reliance on debt to finance middle-class lifestyles. In the middle years of the 20th century, Americans were net savers, putting away as much as 20 percent of earnings for future spending. By the year 2000, Americans had turned into net debtors, falling back on credit cards, mortgage refinancing, and student loans in their efforts to attain the American dream. In her more recent books, such as *Born to Buy* (2004), Schor has drawn attention to how socioeconomic inequality can exacerbate status uncertainty and thus fuel even further debt-financed consumption. The distorted presentation of upper-middle-class lifestyles in the media as the norm and the resultant aspiration to live like those with higher status than ourselves combine to make it all the more likely we will try to live beyond our means, prudent financial planning and concerns for the environment be damned.

The distorting sway of media consumption

In a way, Schor's empirical work on consumerism reveals to be true what French philosopher and cultural critic Guy Debord theorized back in the 1960s. Namely, in the media age, there is a general tendency to grasp the measure of our lives through what we see on television, in the movies and now, of course, through comparative reference to the people and lifestyles we observe online. However, these are not, as Debord would remind us, authentic human experiences. Instead, we find that the lived experience of social interaction – i.e., actual face-to-face relationships with other people – has been colonized by consumer goods and service intermediaries. We are drawn in, for example, by the convenience of "interfacing" with others through online media platforms and social networks, while tacitly accepting the fact that every link we take, every meme we "like" is being tracked for marketing and monetarization. We live, as Debord would put it, under the illusion of the "spectacle," where mediated experience is valorized – both culturally and financially – above direct, non-commodified interaction with other humans (Debord, 1992).

But, "so what?"; you might ask yourself. Perhaps life in the spectacle is simply better than the drudgery of the day-to-day. There are certainly worse things we could be doing with our lives than sitting in front of a screen. In *Born to Buy*, Schor provides sobering survey evidence regarding the impact of media exposure on both the consumer habits and mental well-being of children enrolled at two middle schools in the Boston area. The results are not encouraging. Greater exposure to television and other media was positively associated with greater consumer involvement and negatively associated with children's mental health, measured in terms, of anxiety, depression, self-esteem, and psychosomatic symptoms such as an upset stomach. As mentioned in the chapter on socialization, more recent studies have found similar associations between smartphone usage and feelings of anxiety, loneliness and depression, and sleep deprivation. Excessive time spent on electronic media does appear to have a strong correlation with people's – especially children's – physical and mental well-being.

Re-centering social interaction, generating sustainable practices

This all seems very dark and gloomy, but perhaps there is something to be learned from the inverse of these bleak examples of media overload and alienated ennui. Is there, for instance, any evidence that when people spend more time interacting with each other they are better-off, consume less, and are more willing, perhaps, to engage in ecologically sustainable behaviors? Do our positive connections to others affect, along with our well-being, our willingness to take actions that might reduce our impact on the natural environment? Early work on this topic found a greater willingness on the part of individuals living within supportive communities to consume less during times of crisis (Dietz & Vine, 1982; Heberlein & Warriner, 1983). Deliberate efforts to promote community and conservation, however, often meet with mixed results. The social embeddedness of consumption and the well-funded marketing of highly consumptive lifestyles have proven to be major obstacles in the way of socioecological change (Lutzenhiser, 2002).

Technological efficiencies are never enough

Perhaps more troubling, researchers seeking a link between context and conservation have long observed a "rebound effect" wherein providing people greater access to energy efficient technologies, for example, simply results in them consuming even more (Clark & Foster, 2001; York, Rosa, & Dietz, 2003). In the case of households, greater efficiency and better insulation led to a 31 percent reduction in the energy required to heat the average American home between 1978 and 2005. During that same period, however, the use of electronic appliances nearly doubled and the use of air-conditioning units nearly tripled, essentially canceling out household technological gains in energy efficiency (USEIA, 2011). To put it more directly, what ecological benefit is to be had if you own two or three Energy Star-rated appliances instead of just one?

On the bright side, one positive development in energy conservation has occurred within a practice long considered unchangeable in American culture – automobile usage. Americans on average, and especially younger demographic blocks, now drive less than they did a decade ago, a trend that began before the onset of the 2008 economic downturn and without the encouragement of a coordinated media campaign for curtailment (Dutzik & Baxandall, 2013). Instead, a mix of higher gas prices, lower car sales, and shifts in labor force participation all contributed to what appears to be a lasting change in the nation's driving behavior. There is also considerable evidence that younger generations of Americans are simply more aware of and willing to do something about their ecological footprint (Ibid.). The failure to date of technological efficiencies to bring about a net reduction in global CO_2 emissions combined with the recent reductions in total US miles driven draw our attention to the socially based motivations people have to reduce their environmental impact through conservation.

Along those lines, research on water conservation in Australia has demonstrated that greater connections to neighbors versus friends and family are associated with a greater willingness to participate in conservation-related behavior (Miller & Buys, 2008). Relatedly, greater participation in community activities also correlates with driving less among US commuters (Putnam, 2000; Williamson, 2002). There is thus reason to believe that the exposure of individuals to social interactions grounded in their own neighborhood increases the likelihood of them taking up voluntary practices of conservation that benefit the environment.

How does social context influence our concern for the environment?

My colleague Kristin Williams and I set out to explore further this question of the influence of neighbors on sustainable practices using data from the 2010 General Social Survey (GSS) (Macias & Williams, 2016). We took particular advantage of 80 questions included in that year's "environmental module" addressing established areas of interest within environmental sociology (Stern, Dietz, Abel, Guagnano, & Kalof, 1999), including consumer behavior, a willingness to sacrifice for the environment, environmental activism, and an array of topics tied to environmental values. A key distinction in our study from earlier work is that we brought together environmental attitude questions with seven social capital questions concerning face-to-face interactions with friends, family, and neighbors, community participation, and generalized trust. Control variables in our models include age, gender, race, and ethnic categories, labor market participation, education, household income, urban residence, and political views.

In our statistical analysis, we examined the relationship social capital has with consumer choices, a willingness to make personal sacrifices for the environment, and environmental activism outcomes as measured in the GSS. Among the variables we used to measure social capital, social evenings with neighbors stood out. This proved to be a statistically significant predictor for four of six lifestyle variables – purchasing chemical-free fruits and vegetables, household water conservation, household energy conservation, and driving less. In fact among all our models, social evenings with neighbors was the most frequently significant social capital predictor of pro-environmental behaviors. Why would this be the case?

In social psychology, a "norm activation perspective" would draw our attention to the normative context that encourages environmentally beneficial behaviors as well as the practical barriers in the way of engaging in such behaviors. For example, people whose electric bills include information about average household consumption in their neighborhood are much more likely to take steps to reduce their own energy consumption than people who lack this comparative frame of reference (Schultz, Nolan, Cialdini, Goldstein, & Griskevicius, 2007). Though that example evokes the sway of social context on pro-environmental behavior, it in fact describes a behavior dependent almost entirely on the *perception* of what the neighbors are doing and not necessarily or not at all on social interaction. By introducing social capital into the mix, however, we open the door to the interactive influences social context may have on conservation, namely – reliable sources of information, opportunity, and example.

Why do neighbors matter?

Reliable sources of information

As regards information, the underlying premise is the following: When community members of roughly equal social status have frequent interactions they are likely to create networks of trust and engagement (Bridger & Luloff, 2001). These place-based networks can themselves become reliable sources of information, especially concerning local issues within the community. To be clear, it is not that frequent interactions magically guarantee trust among individuals. It is only through repeated interactions, rather, that one can gain a sense of who has reliable information and who does not; who is trustworthy and who is, perhaps, less so. Regarding conservation, as the cost of energy increases and as household budgets become strained, we may expect that low-cost alternatives in transportation and energy-related strategies for lowering household expenditures will enter the flow of neighborly conversations.

Proximate opportunities to share resources

Beyond simply talking about conservation, however, regular interactions with those nearby create opportunities to share resources and low-impact alternatives to the status quo. It is this particular characteristic of neighborly neighbors – social interactions grounded in place – that appears especially amenable to conservation and curtailment through the sharing of common resources and place-specific information. Community gardens, babysitting circles, and carpooling all represent forms of social organization through which local economies of trust effectively undergird the basis for shared communal resources (Brunie, 2009; Macias, 2008; Newton, 1997). A longstanding institution in American life that should definitely be included here is the public library – a shining case and point of communal resource sharing whose model now extends to more broadly defined lending libraries that include for-loan media equipment, gardening supplies, cooking appliances, and musical instruments (Klinenberg, 2018).

Examples of conservation in an otherwise consumer-oriented culture

Public libraries notwithstanding, one of the biggest challenges in trying to promote conservation in the United States is the dearth of mainstream models to follow. Elected officials are often reluctant to be associated with policies or a message that would encourage people to consume less. Moreover, the central goal of commercial advertising – ubiquitous in the contemporary geographic, electronic, and social landscape – is to promote greater consumption, either of things people already consume or of new products for which demand did not previously exist. Neighborly neighbors thus present a potentially important, if seemingly innocuous, source of conservation through sharing and conversation otherwise unavailable in the dominant culture of electronic media, politics, and commerce.

Jimmy Carter, perhaps the most conservation-minded US President ever elected to the office, encountered this conundrum firsthand. Two weeks into his presidency in 1977 at the height of that decade's energy crisis, he sat before the American public on national television in a beige cardigan sweater and asked them to consume less energy. A few months later he framed his national call for energy conservation as "the moral equivalent of war," imploring his fellow Americans to drive smaller cars, adjust their thermostats, and embrace alternative forms of energy production. This highly unpopular request was not the only reason Carter lost his bid for reelection in 1980 – a US hostage crisis in Iran certainly did not help – but that was the last time a national leader in the United States ever made an open argument for conservation in front of the American people. A few have asked for and even subsidized our consumption of more energy efficient products – see President Obama's 'Cash for Clunker' program, 2009 – but Carter stands out as the conservation president.

Thus, in the current environment, backyard conversations, sidewalk exchanges, and neighborly visits may be some of the best sources of learning about water catchment systems for gardening, carpooling opportunities, savings accrued through thermostat adjustments, and other environmentally friendly practices such as using the public library. Moreover, the structural position of neighbors as near status equals, financially independent of each other make them a key potential source of mutual influence in the realm of conservation. All other things being equal, in a neighborhood with little neighborly interaction we would expect few opportunities for the sort of sharing described here to take place.

Conservation and trust: it's a social thing

Our study, based on a representative sample of the United States, demonstrates that certain interactive social contexts are more strongly associated with conservation and support for pro-environmental policy than others. When we shift the focus away from individual behaviors to the interactive social context of conservation, potential synergies emerge. In transportation, for example, both shorter commutes and lower rates of solo-commuting have been tied to greater participation in community activities (Putnam, 2000; Williamson, 2002). Likewise, public amenities such as benches, community gardens, traffic calming devices, and children's playgrounds have all been shown to increase social interaction within neighborhoods, regardless of the social economic status of the people who live there (Semenza & March, 2009; Wilkerson, Carlson, Yen, & Michael, 2012; Wood, Giles-Corti, Zubrick, & Bulsara, 2013). The movement toward locally-sourced food evident in the growing success of regional farmers markets and community-supported agriculture also points in the direction of greater social interaction, though the question of food equity and affordability continues to be a challenge (La Trobe & Acott, 2000; Lyson & Guptill, 2004; Macias, 2008).

Bewildering rifts within industrial capitalism

One might reasonably ask, however, will spending more time with neighbors and changing our consumer habits really be enough to address the massive ecological crisis we face as living beings on Earth? The answer, of course, is probably not; especially if we fail to address the production side of industrial capitalism; specifically, the imbalance between the quantity of natural resources we exploit and nature's ability to recover from our unchecked plundering.

We began this chapter considering Marx's provocative notion of commodity fetishism; a rather straightforward way to frame how we ignore social costs and human relationships when our attention is focused solely on the consumer object of our desire – all in all a pretty powerful critique of consumer capitalism. Many environmentally-minded thinkers, however, have been left unimpressed with Marx's thinking on ecological issues. Many have argued that by focusing on the "means of production" and labor as the source of value in all things produced through capitalism Marx neglected environmental impacts altogether. One notable exception to this critique has been that of American sociologist John Bellamy Foster. In his book, *Marx's Ecology*, Foster (2000) argues that a central element of Marx's theory has been long ignored; an element more profoundly influenced by the empirical work of British naturalist Charles Darwin than previously recognized.

Metabolic rift is the term used by Foster to describe the central disconnect observed by Marx within industrial capitalism – namely, that between the production of commodities and the disposal of waste. This was reflected most clearly for Marx in the division between the city and the country, a kind of ecological parallel to the conceptual rift between consumer objects and their social origins implied by commodity fetishism. To understand what is meant by metabolic rift, we need to think back to a time when the physical distance between where people produced things and where they consumed things was not very far. Marx points to medieval European villages, but one could include here the *ejido* system in Mexico and any number of other traditional systems of communal land management around the world. In these contexts, villagers work together to cultivate the land and manage collectively provisions for the local population. To speak of a product and a consumer in this context is a bit anachronistic. Commodity fetishism would be almost inconceivable here since the person who ate, used, or otherwise consumed the result of small-scale communal production likely had a hand in its creation. Presumably, everyone would know the social origins of communally produced goods.

Equally important in this metabolic loop, however, were the things left over after they were used – corn cobs, fish bones, broken pottery, spent stalks and stems and, everywhere, piles of poop, human and non-human origin, alike. Kind of disgusting, but also, among a small settlement of a few hundred people, or so, pretty manageable. In fact, in the closed metabolic loop described by Marx, all this refuse (what we would tellingly call today "waste" or "garbage") was essential to the sustained existence of small-scale agricultural settlements. Given the short distances between consumption and agricultural production, leftover material was easily reintegrated back into the soil, providing essential chemical nutrients for life-sustaining agricultural crops.

This all changed with the arrival of industrial capitalism and the concentration of large populations (typically workers seeking wage employment in factories) in towns and cities. Beyond the psychic toll of being socially detached from their places of origin, workers were now no longer directly connected to the fruits of their labor. Their employment in industry was to produce commodities intended for mass distribution. Meanwhile, items workers and their families required for their own sustenance were often produced far from where they lived. On the other end of this socioecological metabolism, refuse that was once reintegrated back into the soil was now redirected into the "waste stream." What this meant practically in many urban households was a bottom floor dedicated to the storage of human excrement, a situation described comprehensively in Steven Johnson's book, *The Ghost Map* (2006). There in mid-19th-century London – not far, incidentally, from where Marx lived the last decades of his life – Johnson provides an account of how early observational research in epidemiology led to the discovery of the root cause of two cholera outbreaks that killed nearly 100,000 of that city's inhabitants. Namely, wastewater had contaminated one of the well pumps for potable water near the city center. There was simply not sufficient infrastructure available at the time to manage the city's mass of raw sewage, most of which ended up ultimately in the Thames river.

Desperate times require desperate measures

This is precisely the "rift" described by Foster in Marx's social metabolism. Not only did the waste produced through industrial capitalism present an ongoing threat to public health, but the soils once replenished through locally sourced organic refuse, were becoming void of plant nutrients. By the mid-19th century, in fact, European and US agricultural producers were experiencing a major crisis of soil depletion, unsure of how to nourish burgeoning urban populations with the increasingly meager output of rural farms (Clark & Foster, 2013). In Europe, battlefields – ad hoc cemeteries, really – from just a few decades earlier were seen as viable sites for extracting nutrient rich soil. The Peruvian economy prospered at this time largely due to the go-go market for bird crap, which it sold as fertilizer in enormous quantities to European countries keen on caca. Out of desperation, the US Congress passed the Guano Act of 1856 through which it seized 94 South Pacific islands solely for their plentiful stock of chiropteran scat. To describe US legislators at the time as "bat shit crazy," would not be entirely inaccurate.

If this were merely a matter of how we managed our manure we could wash the muck off our hands and leave it at that. Alas, it is not. The advent of chemical fertilizer, water treatment plants, and the construction of vast networks of underground sewers in the 20th century did not lessen the rift in our socioecological metabolism. In many ways, they have made it much worse. Though it is a story told and retold now many times over, it is worth underscoring the revolution in industrial agriculture sparked by German chemist and Nobel Prize winner Fritz Haber in the early part of the 20th century. In 1909, Haber found a chemical procedure – the "Haber Process" – for using methane to fix nitrogen into ammonium nitrate, a chemical useful as synthetic fertilizer in agriculture (Erisman, Sutton, Galloway, Klimont, & Winiwarter, 2008).

Given its continuing role in undergirding industrial agriculture, Canadian geographer and policy analyst Vaclav Smil once referred to the Haber process as the "detonator of the population explosion," as it made possible the quadrupling of the human population during the course of the 20th century (Smil, 1999). As Smil well knows, this is an unfortunate if appropriate metaphor since the other major use of ammonium nitrate is in the production of explosives: During times of war, for use in industry, and in domestic terrorist attacks such as the Oklahoma City Bombing in 1995. More recently, the August 4, 2020 devastation of Beirut, Lebanon appears to be the result of the improper storage of nearly 3,000 tons of ammonium nitrate in that city's industrial port (Hubbard et al., 2020).

If the Haber process allowed us as a species to narrowly escape the pending threat of soil depletion, it came at the cost of exponential demographic growth and an ever greater dependency on fossil fuels and their attendant contribution to greenhouse gas emissions. That, as it happens, is only half of Haber's century-long impact on the world's ecosystem. Along with enabling the mass production of explosives, Haber's other major contribution to modern warfare was the fabrication of nerve gas, used to horrific effect in the trenches and on the battlefields of World War I. The chemical legacy of nerve agents developed by Haber and other chemists of that era is evident in many modern pesticides which function essentially as nerve gas for bugs. Again, this was a major advance for increasing crop yields, especially as the mass production of monocrops made plant varieties all the more susceptible to pests and blight. Pesticides (and herbicides) have thus held natural invaders at bay while simultaneously introducing large quantities of fossil fuel-based synthetic chemicals into the environment that were never there before. Currently, the evidence that pesticide exposure is a major factor behind the dramatic decline in insect populations worldwide – including most notably the honeybee and other species essential for pollination – is all but irrefutable (Goulson, 2019).

Increased rationalization of industrial agricultural, however, did not begin and end with Fritz Haber. In the post-World War II era, the Green Revolution – funded by the Rockefeller Foundation and spearheaded by another Nobel Laureate, Norman Borlaug – married chemistry with mechanization, harnessing fossil fuel energy in the use of heavy machinery and the construction of massive feats of civil engineering (Swaminathan, 2009). These included damming rivers, pumping water uphill, and flooding deserts for the irrigation of lands never before considered arable. At this point, agriculture had truly arrived as an industry, driven by the need for growth, economies of scale, and lower production costs. Other developments such as crop hybridization and genetically modified organisms are really just markers along the broad trajectory of rationalization in this core sector of the economy. With every advance, however, the rift between what we produce and its source in nature seems all the more abysmal.

The rift between us and the meat most of us eat

If we shift our gaze slightly towards livestock production and our consumption of meat, the question of metabolic rift comes into even sharper relief. Never before in the history of humankind have so many people eaten so much meat. You may, for example, in your circle of family, friends, and acquaintances know a number of people (including yourself, perhaps) who are vegetarians or vegans. These numbers, however, pale in comparison to the tsunami of carnivorous homo sapiens whose demand for animal flesh has skyrocketed over recent decades. The factory production of meat is a major global industry with profound environmental impacts, and meat eating among humans today is closely tied to higher levels of income (Sans & Combris, 2015). As the middle class has expanded around the world, so has the taste for carne asada and filet mignon.

Consistent with our riff on rift, the slaughtering and processing of all this meat happens far out of view from most consumers, in abattoirs and meat-packing plants that, thankfully, most of us will never have to work in. Industrial meat-processing consistently ranks as the most dangerous occupation in the United States with about 30 percent of workers in this sector experiencing a work-sustained injury annually, including lacerations, torn ligaments, broken bones, and amputations (Schlosser, 2012). Given its low level of desirability among domestic workers, it is not surprising that this industry is heavily dependent on immigrants and undocumented labor.

Though an obvious point, it is not like the animals have it any better. Industrial beef, pork, and poultry production is about as close an approximation to Dante Alighieri's account of the ninth circle of hell that you will find on Earth. A confined existence immersed in feces and punctuated by mutilation, self-mutilation, and the constant threat of infection would be a succinct way of describing the short, cruel life of a factory farm animal. The threat of infection has implications far beyond the filthy stockyards and grimy cages these creatures of the damned inhabit. The industry, which has a vested interest in their livestock not dying before they reach the slaughterhouse, has found the ultimate solution to this problem – antibiotics.

You may have read or seen reports about the overuse of antibiotics among the human population: Physicians prescribing penicillin for a viral throat infection, for example, even though antibiotics work on bacterial infections and have no effect whatsoever on viruses. You may have also learned that many hospitals are at a point of crisis because of the antibiotic resistance microbes have built up as a direct result of over-prescription. You may be surprised, however, to find out that the amount of antibiotics used in the industrial production of pork, beef, and poultry is about four times as much as that used on humans in the United States every year (Martin, Thottathil, & Newman, 2015). In fact, if we were really concerned about the rise of antibiotic resistant strains of bacteria, we would probably do much better by focusing our efforts on reducing the overuse of these chemicals on livestock rather than people. So, what is going on here? How did pharmaceuticals end up in our hamburgers, hot dogs, and chicken nuggets?

There are two reasons for this. The first is rather obvious – as mentioned above, the heinous conditions through which factory farm animals live out their lives places them in constant risk of infection. Instead of improving living conditions, industrial meat producers have simply opted to preemptively inoculate their product on a regular basis with antibiotics. The second reason is actually more of a side benefit – though it is still not entirely clear why, antibiotics can also shorten the time it takes for animals to reach maturity, and the sooner an animal matures, the sooner it can be taken to market for slaughter; a great boon for the producer. For consumers, this means our global demand for cheap meat is readily met. The environmental impact, however, is enormous.

Consider these helpful bullet points from the United Nations (UN) publication, 'Livestock's Long Shadow' (Steinfeld et al., 2006):

- Animal livestock is responsible for 18 percent of greenhouse gas emissions, more than transportation
- Meat production uses 8 percent of the world's fresh water supply, mainly for feed crops
- Twenty-six percent of the earth's land surface is used for livestock production
- Seventy percent of previously forested land in the Amazon is now used for grazing
- Factory farm runoff – also laced with pharmaceuticals, hormones and food crop chemicals – is a major source of water contamination in the world, today

To be clear, the UN is not known for its radical stance on politically unpopular issues. Nonetheless, a mainstream global institution whose primary goals center around assuring peace, security, and the avoidance of war, is telling most people on earth that our appetite for animal protein – more so than our addiction to driving – is killing the planet, and we might want to consider this the next time we shop for groceries. Sadly, since the publication of this UN report, all indications are that the predictive link between disposable income and meat consumption has not been broken. Statistically speaking, most people only reduce their meat consumption when they can no longer afford it. Individual metabolisms trump the socioecological metabolism at every turn, and informing people about their ecological impact appears to sway only a very small percentage of the population.

The connection between our own metabolisms and the broader socioecological metabolism speaks to our biology as living things whose urges and drives evolved when there was a relative paucity of resources available in the natural world. In pre-industrial times, to be hungry was to have the will to persevere and survive; we consume things like sugar and meat because our bodies crave them. Sure, our bodies and minds can be socialized into having more refined tastes, but ultimately we need energy and protein to live and it should come as no surprise that we are hardwired to enjoy food that can provide us these things when they are available.

Metabolic rift is thus another way of describing how industrial capitalism, through its harnessing of fossil fuels, has released us from the shackles of scarcity, allowing our biologically based desires and cravings to take precedence over long-term planning and addressing broad collective action problems such as climate change or social inequality. With the right level of income, you too may attain that light-as-air sensation that you are economically autonomous, dependent on nothing but yourself for success or failure in the world.

The problem of late, as you may have noticed, is that the rift between our modern lifestyles and expectations and the natural world that makes them possible is getting too big to ignore. Hotter temperatures, year-round forest fires, an extended hurricane season, and coastal flooding are all part of a global pattern of increased weather volatility. Whatever remains of political resistance to the science of climate change, governments around the world cannot ignore this mounting threat to their citizens and regional economies. Even in the United States where some political leaders express public doubt about climate science, the military – i.e., the largest category of discretionary spending on the national budget – has declared that climate change poses a threat to national security. A 2019 Pentagon report identified climate-based risks in 79 US facilities around the world due to severe weather and flooding. This same report warned of greater levels of political instability in resource-threatened regions, and signaled the need for greater humanitarian assistance in places impacted by climate volatility (Collins, 2019).

The farcical trap that is not so funny

The rift is real but, up to now, the dominant strategy for addressing our socio-metabolic imbalance has been to find a synthetic workaround. Soil depletion? Inject the ground with fossil-fuel sourced fertilizers. Limited varieties of agricultural crops become highly susceptible to pests and blight? Spray them with chemical pesticides and herbicides. Animal livestock living in heinous conditions are under constant threat of infection? Douse them in antibiotics. And the list goes on. In their article, 'Tragedy of the Commodity: The Overexploitation of the Mediterranean Bluefin Tuna Fishery,' Stefano Longo and Rebecca Causen bring up one of Marx's more famous adages – "history repeats itself…, first as tragedy, than as farce" – to highlight the way capitalism often addresses environmental problems (Longo & Clausen, 2011). In the case of Mediterranean bluefin tuna, the tragedy of overfishing was met with the farce of tuna "ranching" and aquaculture; both of which intensify Bluefin tuna production yet do nothing to address the underlying issue of overexploitation.

The "tragedy of the commodity" is in fact a critique of the "tragedy of the commons," a theory originally forwarded by Garrett Hardin (1968), which argues that, left to their own devices, people living off a common shared resource – shepherds and their pastures, for example – will naturally tend to overexploit that resource following a selfish, competitive impulse. In contrast, Longo and Causen draw our attention to the way commodity capitalism transforms long-standing traditions of shared resource management – Sicilian trap fisheries, in their work – into strategies of extraction which view nature as merely a material input along the path to greater financial yield. To recap the farcical side of tech solutions for ecological troubles: Along with requiring greater capital inputs to pull off, they often conveniently avoid the question of resource conservation, while – as you would expect of any capital investment – working primarily as a catalyst for short-term profit.

Sadly, the single most popular sustainable activity that Americans engage in – recycling – also falls into this pattern of tragedy and farce. The tragedy, in this case, was represented in dramatic fashion by an early 1970s public service announcement calling attention to the increasingly polluted highways and byways of the American landscape. In it, a Native American played by Espera Oscar de Corti – an Italian-American actor who preferred the pseudonym "Iron Eyes Cody" – is shown paddling a canoe out of pristine nature into a polluted river, ultimately arriving at the outskirts of an American city full of smokestacks and overrun with automobiles. He steps cautiously over trash strewn on the roadside when someone in a passing car tosses a bag of garbage at his feet. The camera then zooms in on the distraught Indian, closing in on his eyes, and then a single tear running down his cheek. Bracketing for the moment the stereotypical depiction of an American Indian by a European American actor, the portrayal of the polluted landscape was accurate and captured the increasingly contaminated environment which, by the early seventies, had gained traction as a political issue, resonant enough to move Congress to pass a series of environmental protection laws during that decade.

The farce, however, comes at the end of the ad sponsored by the Keep America Beautiful campaign which declares, "People start pollution, people can stop it." "What's so farcical about that?" you might ask. Context, however, is anything. As clearly laid out in Heather Roger's book, *Gone Tomorrow: The Hidden Life of Garbage (2006)*, the primary sponsors of the KAB Campaign have been the bottling industry which for over 60 years has been trying to avoid responsibility for the mountains of garbage they produce every day. It all started back in 1953 when my home state of Vermont successfully passed a bottle bill that outlawed the sale of non-refillable containers. On the face of it, this was not a big deal – even I growing up in the 1970s remember my parents returning used cola and beer bottles back to the store they were bought for a deposit.

Vermont legislators, however, saw the future; specifically, industry's growing preference for single-use containers. If Coca-Cola, Anheuser-Busch, and other bottlers could just convince Americans that consumers were solely responsible for the disposal of their containers, they could save tons of money. By the time of the KAB ad, many other states had passed their own bottle bills, a highly effective strategy for reducing litter – states with bottle deposit laws have on average a 60 percent container recovery rate compared to 24 percent in states without them (Gitlitz, 2013). Seeing a looming threat to profit, industry ramped up pressure on state legislators against bottle bills, and teamed up with the Ad Council to produce iconic advertising against government regulation.

Cynically tapping into gnawing guilt about the historical treatment of American Indians and the public's growing awareness of environmental problems, the KAB campaign proved a huge success, marking as it did a monumental shift of container responsibility away from industry to the consumer. No longer would bottlers have to cover the costs of recycling, washing, and relabeling used containers. Instead, it was consumers who would have to sort and figure out what to do with the growing universe of plastic containers, most of which cannot be reused directly, and much of which ends up degraded in the world's oceans where it is estimated that by 2050 there will be more plastic by mass than fish (Neufeld, Stassen, Sheppard, & Gilman, 2016). Plastic recycling, as it is currently carried out, is a global farce of misdirection, imploring the consumer to accomplish what only a regulated industry is capable of; namely, bringing an end to single-use plastic containers.

In this chapter, we have considered three enormous disconnects characteristic of our place in human history. First, commodity fetishism points us to the way in which we neglect the social origins of almost everything we consume. As discussed above, both producers and consumers are to some degree motivated to ignore the less savory aspects of commodity production. Be it on the assembly lines of fast fashion apparel or the industrial farms and abattoirs that most of us rely on for sustenance; everything seems to go down better if we do not know exactly how it was produced. Second, the metabolic rift reminds us of the great ecological imbalance and environmental costs inherent in industrial capitalism where, not only are resources extracted at an accelerated pace, but the waste produced through production becomes an additional problem in need of a solution which capitalists cannot afford to pay for. Third, the solutions that are proposed to address environmental problems caused by capitalism are often farcical; a Band-Aid, or distraction from the actual problem whose actual solution is usually conservation, that is, the reduced consumption of a scarce resource.

Glossary

Commodity Fetishism   Karl Marx's term to referring to the pattern in capitalist societies of gravitating towards the physical qualities of a good without taking into consideration its social origins in human labor.

Conspicuous Consumption   Use of unproductive goods and experiences with the primary goal of conveying superior socioeconomic status to others.

Conspicuous Leisure   The act of unproductive labor for social recognition.

Green Revolution   The post-WWII period of technological advancements that mechanized agriculture.

Haber Process   The Nobel Prize-winning technique developed by Fritz Haber that synthesized ammonium nitrate for fertilizer, thus underpinning the growth of industrial agriculture during the 20th century.

Metabolic Rift   John Bellamy Foster's Marxian idea capturing the environmental and social disconnect between industrial production and the waste produced.

Rebound Effect   The tendency for industry and consumers when presented with energy efficient technologies to simply increase the utilization of those technologies, thus cancelling out potential gains in energy conservation.

Spectacle   The contemporary lens of analyzing and tweaking our own lives through comparisons with representations on TV, film, and social media; an idea developed by cultural theorist Guy Debord.

Tragedy of the Commodity   The tendency within capitalism to address environmental problems with distractions requiring higher levels of consumption, when a more obvious solution is conservation; that is, the reduced consumption of a scarce resource.

Vicarious Consumption   The ability of one to utilize someone else's economic power – often a family member – to display wealth and status.

Questions

1   Do you have a commodity fetish? Think of your favorite consumer object that is not a smartphone and write down all its features and characteristics that bring you satisfaction. Now, get online and do some research. Where was this thing made? Under what conditions and in which country? Are workers there allowed to organize? Where did the material resources necessary to make it come from? What are the environmental consequences of its production? Are you happy, now?!? Seriously, do you think having a greater awareness of how commodities are produced would result in our reduced consumption of them?

2   What is your relationship with waste? What changes have you seen in recent years that give you hope people are becoming more thoughtful about how they dispose of refuse? What role have government and local institutions played in bringing about these changes? Some communities and individuals now advocate for a shift to "zero waste." Is that even possible? Do a little research on the topic. Is it something you would consider trying?

3   Plastic bottle "recycling," aquaculture, and much of industrial agriculture are pointed to in this chapter as examples of farcical solutions to

mounting ecological crises. Can you think of other farcical superficial solutions to underlying socioecological problems in modern life? Who benefits from these misdirected strategies? What can be done to reveal the true motives behind them and change course away from short-term profits, towards long-term, sustainable planning?

References

Bourdieu, P. (1984). *Distinction: A social critique of the judgement of taste*: Harvard University Press.

Bourdieu, P. (2011). The forms of capital (1986). *Cultural Theory: An Anthology, 1*, 81–93.

Bridger, J. C., & Luloff, A. E. (2001). Building the sustainable community: Is social capital the answer? *Sociological Inquiry, 71*(4), 458–472.

Brunie, A. (2009). Meaningful distinctions within a concept: Relational, collective, and generalized social capital. *Social Science Research, 38*(2), 251–265.

Clark, B., & Foster, J. B. (2001). William Stanley Jevons and the coal question: An introduction to Jevons's "Of the Economy of Fuel". *Organization & Environment, 14*(1), 93–98.

Clark, B., & Foster, J. B. (2013). Guano: The global metabolic rift and the fertilizer trade. In A. Hornborg, B. Clark, and K. Hermele (Eds.), *Ecology and power* (pp. 84–98): Routledge.

Collins, C. (2019). *Climate change and global security: Planning for potential conflict*: Climate Institute, Washington, D.C.

Debord, G. (1992). *The society of the spectacle 1967*: Les Éditions Gallimard.

Dietz, T., & Vine, E. L. (1982). Energy impacts of a municipal conservation policy. *Energy, 7*(9), 755–758.

Dutzik, T., & Baxandall, P. (2013). *A new direction: Our changing relationship with driving and the implications for America's future*: U.S. Public Interest Research Group Education Fund.

Erisman, J. W., Sutton, M. A., Galloway, J., Klimont, Z., & Winiwarter, W. (2008). How a century of ammonia synthesis changed the world. *Nature Geoscience, 1*(10), 636.

Foster, J. B. (2000). *Marx's ecology: Materialism and nature*: New York University Press.

Gitlitz, J. (2013). *Bottled up: Beverage container recycling stagnates (2000–2010)*: Container Recycling Institute.

Goulson, D. (2019). The insect apocalypse, and why it matters. *Current Biology, 29*(19), R967–R971.

Gunders, D., & Bloom, J. (2017). *Wasted: How America is losing up to 40 percent of its food from farm to fork to landfill*: Natural Resources Defense Council New York.

Hardin, G. (1968). The tragedy of the commons. *Science, 162*(3859), 1243–1248.

Heberlein, T. A., & Warriner, G. K. (1983). The influence of price and attitude on shifting residential electricity consumption from on-to off-peak periods. *Journal of Economic Psychology, 4*(1–2), 107–130.

Hochschild, A. R. (1997). *The time bind: When work becomes home and home becomes work*: Macmillan.

Hubbard, B., Abi-Habib, M., El-Naggar, M., McCann, A., Singhvi, A., Glanz, J., & White, J. (2020). How a Massive Bomb Came Together in Beirut's Port. *New York Times (Online)*.

Johnson, S. (2006). *The ghost map: The story of London's most terrifying epidemic—and how it changed science, cities, and the modern world*: Penguin.

Klinenberg, E. (2018). *Palaces for the people: How social infrastructure can help fight inequality, polarization, and the decline of civic life*: Broadway Books.

La Trobe, H. L., & Acott, T. G. (2000). Localising the global food system. *International Journal of Sustainable Development & World Ecology, 7*(4), 309–320.

Longo, S. B., & Clausen, R. (2011). The tragedy of the commodity: The overexploitation of the Mediterranean Bluefin tuna fishery. *Organization & Environment, 24*(3), 312–328.

Lutzenhiser, L. (2002). Marketing household energy conservation: The message and the reality. In T. Deitz, & P. Stern (Eds.), *New Tools for Environmental Protection: Education, Information, and Voluntary Measures* (pp. 49–65): National Academy Press.

Lyson, T. A., & Guptill, A. (2004). Commodity agriculture, civic agriculture and the future of US farming. *Rural Sociology, 69*(3), 370–385.

Macias, T. (2008). Working toward a just, equitable, and local food system: The social impact of community-based agriculture. *Social Science Quarterly, 89*(5), 1086–1101.

Macias, T., & Williams, K. (2016). Know your neighbors, save the planet: Social capital and the widening wedge of pro-environmental outcomes. *Environment and Behavior, 48*(3), 391–420.

Martin, M. J., Thottathil, S. E., & Newman, T. B. (2015). *Antibiotics overuse in animal agriculture: A call to action for health care providers*: American Public Health Association.

Miller, E., & Buys, L. (2008). The impact of social capital on residential water-affecting behaviors in a drought-prone Australian community. *Society and Natural Resources, 21*(3), 244–257.

Neufeld, L., Stassen, F., Sheppard, R., & Gilman, T. (2016). *The new plastics economy: Rethinking the future of plastics.* Paper presented at the World Economic Forum.

Newton, K. (1997). Social capital and democracy. *American Behavioral Scientist, 40*(5), 575–586.

Putnam, R. D. (2000). *Bowling alone: The collapse and revival of American community*: Simon and Schuster.

Rogers, H. (2006). *Gone tomorrow: The hidden life of garbage*: The New Press.

Sans, P., & Combris, P. (2015). World meat consumption patterns: An overview of the last fifty years (1961–2011). *Meat Science, 109*, 106–111.

Schlosser, E. (2012). *Fast food nation: The dark side of the all-American meal*: Houghton Mifflin Harcourt.

Schor, J. (1999). *The Overspent American: Why we want what we don't need*: HarperCollins.

Schor, J. (2004). *Born to buy: The commercialized child and the new consumer culture*: Simon and Schuster.

Schultz, P. W., Nolan, J. M., Cialdini, R. B., Goldstein, N. J., & Griskevicius, V. (2007). The constructive, destructive, and reconstructive power of social norms. *Psychological Science, 18*(5), 429–434.

Semenza, J. C., & March, T. L. (2009). An urban community-based intervention to advance social interactions. *Environment and Behavior, 41*(1), 22–42.

Smil, V. (1999). Detonator of the population explosion. *Nature, 400*(6743), 415.

Steinfeld, H., Gerber, P., Wassenaar, T., Castel, V., Rosales, M., Rosales, M., & de Haan, C. (2006). *Livestock's long shadow: Environmental issues and options*: Food & Agriculture Org.

Stern, P. C., Dietz, T., Abel, T., Guagnano, G. A., & Kalof, L. (1999). A value-belief-norm theory of support for social movements: The case of environmentalism. *Human Ecology Review*, 6(2), 81–97.

Swaminathan, M. (2009). Norman E. Borlaug (1914–2009): Plant scientist who transformed global food production. *Nature*, *461*(7266), 894–895.

USEIA. (2011). *Share of energy used by appliances and consumer electronics increases in U.S. homes*. Retrieved from https://www.eia.gov/consumption/residential/reports/2009/electronics.php. Accessed September 16, 2021.

Veblen, T. (2017). *The theory of the leisure class*: Routledge.

Wilkerson, A., Carlson, N. E., Yen, I. H., & Michael, Y. L. (2012). Neighborhood physical features and relationships with neighbors: Does positive physical environment increase neighborliness? *Environment and Behavior*, *44*(5), 595–615.

Williamson, T. (2002). Sprawl, politics, and participation: A preliminary analysis. *National Civic Review*, *91*(3), 235–244.

Wood, L., Giles-Corti, B., Zubrick, S. R., & Bulsara, M. K. (2013). "Through the Kids... We Connected With Our Community" children as catalysts of social capital. *Environment and Behavior*, *45*(3), 344–368.

York, R., Rosa, E. A., & Dietz, T. (2003). STIRPAT, IPAT and ImPACT: Analytic tools for unpacking the driving forces of environmental impacts. *Ecological Economics*, *46*(3), 351–365.

6 Homophily and the social strictures of an unequal society

Sometimes a single word seems to sum up the mood of the times. In the late-1960s, it might have been "groovy," capturing both the twisty-turny, psychedelic atmosphere and central culture-bearing significance of vinyl records during that era. As an admittedly unhip sociologist, I would like to propose "homophily" (rhymes with falafel-y) as the standout one-word summary of our current day and age. "What the hell," you might ask, "are you talking about?"

Very briefly, homophily refers to the tendency for people to prefer the company of others like themselves. We will spend much of this chapter trying to unpack the significance of that definition for society today. I want to start, however, by pointing out the important implications this term has for both methodology and theory within sociology. On the methodological side, social scientists are often at risk of misreading homophily as something else. For example, in the previous chapter, I cited research my colleague Kristin Williams and I made concerning the relationship between spending time with your neighbors and sustainable behaviors such as water conservation and driving less. We presented generalizable national data that showed a statistically significant relationship between neighborliness and sustainable behaviors.

We were careful, however, not to use the language of causality in our analysis – i.e., we did not say that neighborliness causes sustainable practices among individuals. One good reason for that is homophily – we cannot rule out the possibility that people who like to spend time with their neighbors also just happen to be people who like to engage in sustainable practices. That is, "like attracts like," or more colloquially, "birds of a feather flock together." As opposed to indicating a causal relationship, our research might simply be pointing to the fact that, in modern life, people with similar characteristics are often drawn to each other (McPherson, Smith-Lovin, & Cook, 2001). You could even imagine, if you were so inclined, an exaggerated stereotype of political liberals who believe in communal living – i.e., spending lots of time with their neighbors – and who also take targeted actions to save the planet such as riding their bike to work, growing their own vegetables, and sharing communal resources. In that scenario, the causal force behind

DOI: 10.4324/9781003110668-6

behavior may instead be – as opposed to our explanation – an elevated awareness of one's own ecological footprint, influencing both neighborliness and sustainable practices.

In our defense, we are not the only social scientists facing the charge of possibly misinterpreting our results. Without laboratory research (which is often not generalizable) or logistically challenging panel studies conducted with the same group of people over long periods of time, it is essentially impossible to demonstrate causal relationships with survey data such as those we had at our disposal. We instead do our best to incorporate statistical controls for income, education, political views, gender, and other variables that might intervene between the neighborliness/sustainability relationship. That is, if after including all these controls we find that our relationship of interest persists, we can at least be confident that our principal hypothesis cannot yet be dismissed. This is not a perfect solution, but it does lessen the likelihood that the significant association we find between neighborliness and sustainable behavior is not simply a result of homophilic relations – e.g., likeminded eco-liberals choosing to socialize with others on the basis of education, income, and political views.

Beyond these methodological challenges, how can one argue that homophily is the key underlying tendency in modern life? To be sure, I am not saying our preference for being around people similar to ourselves is necessarily a conscious choice. In fact, though sociologists are interested in this individual-level inclination, the truly interesting theoretical side of homophily is how it can structure our lives in ways we are not even aware. This is closely related to Bourdieu's notion of cultural capital that we referred to in the previous chapter. There, we considered how consumer tastes often track closely with one's social class and education.

On a more interactive basis, however, we can also imagine how our personal stock of cultural capital can work as a kind of sorting mechanism through which we (consciously or not) pick and choose our friends and allies. In an individualistic culture like our own we tend to think of things that make us unique and special relative to the people we know best. If we step back, however, the more remarkable thing is how much we have in common with the people around us. Our favorite television shows, our cultivated knowledge of local cafes and restaurants, the books we read, the sports we play – all of these both define us and link us socially to other people in our lives. Sometimes this is deliberate, as when we attend a theatrical performance principally as a way to impress a romantic interest, for example.

Much of our cultural capital, however, is so deeply ingrained in our being – our manner of speaking, our vocabulary, the way we carry ourselves in different contexts – that Bourdieu even coined a word for it: *habitus* (Bourdieu, 1977). Literally, this refers to the physical embodiment of our social class status. Not only do we convey our habitus to others with our every gesture and utterance, we also subtly perceive it in others – their cultural references, their knowledge of rugby, that interesting accent, those amazing shoes.

Significantly, we may find ourselves drawn to those whose habitus is similar to our own. This, in fact, is one of the key ways homophily operates in contemporary social life. We are not, as it turns out, randomly placed together with our friends and significant others. There is a lot of sorting going on and much of it, as Bourdieu perceived, is on the basis of culture and education. It is still the case that the single best predictor two people are likely to marry each other is their corresponding levels of educational attainment (Musick, Brand, & Davis, 2012). The fact that marriage itself is increasingly a signal of class status with less educated people in the United States being unlikely to marry at all, only underscores this point (Fry, 2014).

Of course, in many ways this cultural and educational sorting is rather innocuous and may even facilitate our ability to interact socially. Our level of comfort is greatly increased in the company of others who (literally and figuratively) speak our language. When I do not have to explain my esoteric references to prog rock (what does "In-A-Gadda-Da-Vida" really mean, anyway?) or French sociology (how would you describe *your* habitus?) the conversation flows a lot more smoothly. For practical reasons, we often seek out people who in a fundamental way can really understand us. Birds of a feather do flock together and, at the psychic level of connection and social interaction, it makes good sense.

The darker side of a deeper connection

There is, however, another side to this dominant tendency in modern life where digital technology, if anything, hones our ability to select people like ourselves with whom we feel most at ease. Who, for example, is being left out of our circle of family and friends? What happens at the societal level when the general tendency is to associate primarily with those who are most like us, economically, demographically, and culturally speaking? Not surprisingly, many social scientists have observed an accelerated sorting of the US population in recent decades (Bottero, 2007; Huber & Malhotra, 2017; McPherson et al., 2001). This is perhaps most obvious in the realm of politics where the geographic distribution of voters has been depicted as the difference between "red" (conservative, middle America) and "blue" (coastal, liberal) states. Though sometimes exaggerated, stark political contrasts between rural and metropolitan areas in the United States are quite clear. When we consider the racial, religious, and ethnic differences built into these divisions, the broader impact of individual preferences on the basis of social comfort appears all the more consequential.

How did this happen; how did we become so divided and compartmentalized as a nation? Sociology, as it happens, is uniquely equipped as a discipline to address this question. Its origins in the 19th century were based on an impulse to better understand the complex interrelations of people living in a transformed industrial landscape as well as the nature of inequality in a context now layered with diverse categories of class and social status.

To understand homophily, it often helps by considering another sociological term – *stratification* – which we shamelessly borrow from our colleagues who study rocks. Admittedly, the only commonality between social stratification and geological stratification is that in both cases researchers are studying layers. In the case of sociology, stratification points to the fact that throughout history we can observe a tendency for hierarchy – i.e., maintaining a ranked order of people in human societies.

A brief history of hierarchy in human societies

When we look at this tendency over time, we see that some societies are more hierarchical than others. Archeological and anthropological research suggests that hunter-gatherer societies have very few layers of social categories and, on the whole, appear to be highly egalitarian, especially as compared to the modern context (Boehm & Boehm, 2009). The underlying factor here appears to be complexity. Small groups of 50 nomadic humans, for example, who migrate to different sites based on the seasons and the availability of edible vegetation, game, and water, do not require multiple layers of social hierarchy. Certainly, in a resource scarce environment, attempts at signaling superior status through conspicuous waste, such as Veblen observed in his *Theory of the Leisure Class* (Veblen, 2017), would not be easily tolerated. The irony is that life as a hunter-gatherer was full of leisure. Anthropological estimates of hunter-gatherer societies that persisted into the 20th century are that the average person labored between 15 and 20 hours a week to do what was needed to survive (Sahlins, 1998). Life in this context was mainly rest and relaxation, colored by an intense engagement with the natural world and an unambiguous interdependence with other human beings.

With the emergence of agriculture and permanent human settlements some eleven thousand years ago, life among Homo sapiens became considerably more complex and just a bit more hierarchical. Making it through the winter or simply adjusting to annual changes in rainfall and water levels, requires planning, cooperation, and some degree of direction. Moreover, the ability to accumulate grain and other agricultural crops introduced the possibility of surplus production. To a greater degree than before, people in human settlements did not have to devote most of their labor to producing food (Lenski, 2013). Cobblers, masons, blacksmiths, and bakers were all specialists who added an additional layer of complexity in villages and townships that were also increasingly in need of leadership to coordinate and manage collective interests. Leaders in this context were usually not selected through a democratic process; more typically leadership was tied to cultural traditions, including religions which could imbue superiors with an otherworldly sanctity that itself represented a significant layer of status above the commoner.

As agriculture became more efficient, human settlements became larger and the stratified layers of social status all the more complex and diverse. Sociologists have highlighted a few dominant systems of stratification that have appeared in different parts of the world to deal with increasingly complex societies. *Slavery* refers to a system within which one dominant group in society literally owns members of another group as property. The master/slave relationship was evident during ancient Egyptian, Greek, and Roman civilizations, and was an inescapable part of life for many people during most of American history, dating back to the colonial period until the official emancipation of slaves at the end of the Civil War in 1865. Slavery, however, persists today, notably in the sex trade where people, often from a very young age, are owned and maintained as property to generate revenue. By some accounts, there are currently more enslaved people on the planet than at any other time in human history (Bales, 2012).

Another form of social stratification which might be characterized as culturally enforced homophily is the *caste system*. In this context, a strict scheme of social norms is in place to maintain cultural boundaries between groups of different social rank. This is often most evident in the marriage market where the nuptial union between certain groups is forbidden. The caste system in India, which officially ended with independence in 1948, is often pointed to as a recent example of this sort of rigid hierarchy. However, "anti-miscegenation" laws restricting interracial marriage in many parts of the country point to our own history of caste in the United States. It was not until the Supreme Court decision of *Loving v. the State of Virginia* in 1967 that such laws – still on the books in 17 states – were finally deemed unconstitutional (Livingston & Brown, 2017). Segregated schools, housing, and other discriminatory practices in the public and private sector have also worked in American culture to maintain a race-based caste system through much of our nation's history.

Class-based stratification

Today, social class status represents the primary basis of social stratification in modern societies. In class-based stratification, economic differences between groups matter most. That may sound a bit harsh but there are certain advantages and liberties that accompany this particular system. In that regard, the most important characteristic of class- based stratification is the notion of "free labor" – unlike slavery, for example, or the medieval feudal system, workers are not bound to a master or lord. We work, instead through a legal agreement with our employer (Engerman, 1999). This state of free labor is so natural to us we rarely give it a second thought.

If you have ever held a job, you may recall the moment not long after your new boss congratulates you, when you must sign the documents that make your employment official. It is then that you freely enter in to a contract with your new employer. If you agree to carry out the job as defined, s/he will pay you on a regular basis the recompense indicated on the contract. What is truly "free" about this situation is that if at any time you feel you no longer want this job – e.g., it pays to little, the hours are horrible, your boss is a tyrant, or you have a better offer somewhere else – you may exit this agreement.

Of course, the same goes for your employers. If they feel you are not carrying out the work you agreed to or for other reasons such as lack of profitability in the business, they may let you go as well. Barring certain legal protections, neither you nor your employer is permanently bound, one to the other.

Our ability to move around and find employment that best fits our qualifications not to mention our financial needs, has enormous implications for the way we experience the trajectory of life. At least theoretically, and in contrast to the caste system, for example, social status is somewhat achieved. As your mother or father probably told you at one point or another, if you work hard and study you can become anything you want. There is, as you have probably figured out by now, an element of truth to that phrase within a class-stratified society. The term sociologists use to describe this possible move up or down the class hierarchy is *social mobility*. In the United States, of course, the focus is usually on the move up, and through the middle part of the 20th century, especially after World War II, there was indeed a general tendency for people to have a higher standard of living than their parents did before them.

As economist Robert Gordon points out, the United States experienced a remarkable period of economic growth between the years 1948 and 1972 (Gordon, 2017). Never-before-seen advances in technology and the industrialization of the economy spurred on in part by the war effort helped generate an era characterized by nonstop economic growth. At the same time, a progressive federal tax system combined with record levels of union membership among workers assured economic gains were redistributed throughout society. In that context, class-based stratification was a godsend for many households who, over generations, could feasibly work their way out of poverty into the middle class as the rising tide of economic growth lifted many, many boats. This form of social mobility highly dependent on the context of rapid growth is known as *structural mobility* – specifically, this refers to a context where the growth of jobs outpaces the numbers of workers available to fill them. In that situation, employers have no choice but to offer higher pay in competition for a limited number of workers.

Measuring socioeconomic status

All this talk of social class stratification, structural mobility, and moving up and down the ranks of a hierarchical society can seem a bit abstract. Now is probably a good to time be more specific about exactly what sociologists are referring to in terms of people's socioeconomic status, also known as, SES. For Karl Marx writing in the 19th century, everything was rather straightforward – people could be divided into two major categories; the profit-seeking capitalists who owned all the factories and "means of production," and the workers who were forced to sell their labor at a level less than its true productive value, thus making profit possible for the capitalist owners. Arguably, things were more complicated even then but, without a doubt, current living and working conditions are dramatically different from those of Europe 200 years ago. Recognizing this complexity, contemporary sociologists have taken four primary strategies for trying to measure differences in SES – household income, wealth, education, and occupational status.

Income

When people think of SES, income is often the first characteristic that comes to mind, and with good reason. The current balance on our bank account is something most of us pay attention to with considerable interest. This determines our ability to cover rent, pay the bills, and eat on a regular basis, not to mention what we are able to do with our free time. Unless someone else is covering all these costs (lucky you!), you probably have a concrete understanding of your personal earnings. The same logic tends to apply in sociological data – income data on working Americans is readily available from a variety of sources. The US government, for example regularly collects information about individual and household earnings, most obviously from annual tax filings. Besides the Internal Revenue Service, the US Census Bureau and the Bureau of Labor Statistics also track income systematically in their survey data, much of which is available for free online. Non-governmental organizations such as the Pew Research Center, the National Opinion Research Center, and many others also collect this kind of information. Thus, in addition to being an easily understood indicator of one's economic standing, income is also a great measure of SES because it is readily available.

That said, income does pose some serious problems if used as the sole measure of SES. Geography, for example, can greatly distort the significance of a specific level of income in different parts of the country. You may have noticed that the cost of living in certain parts of the country – especially near the coasts – is a good deal more expensive than in other parts. For the rent you pay to live in a tiny apartment in the Park Slope neighborhood of Brooklyn, or San Francisco's Mission District, you could easily make mortgage payments on a spacious home in Cleveland, Ohio, or Tulsa, Oklahoma. As a comparative measure across the diverse landscape of the United States, income falls short. Even within the same neighborhood, measures of individual income may be deceptive. For this reason, when comparing SES data, sociologists usually look at household income, recognizing that your financial status in life is greatly impacted by the collective income of a household. The differences between the same individual income for a single parent with three children versus a partner in a two-income household with no kids are enormous.

Wealth

Beyond the differential meaning of the same income within and across households, there is another major determinant of SES which is even more difficult to measure. If you have ever visited places such as Brooklyn and the Bay Area, California, or Washington, D.C., and Seattle, for that matter, you may have wondered, how is it possible that all these twenty-somethings just out of college can afford to live here given their meager incomes? There are at least two answers to this question. First, it is possible that many of these young adults are indeed independently earning enough to afford the cost of living in these cities, as well as service any debt they may have incurred up to that point in their lives. A second possible answer is they are receiving help, as in financial assistance from their family. It is at this point that the question of family wealth emerges as a key factor in the trajectory of an individual's life course in a stratified society.

Of course, the advantages of wealth incur long before a child begins his/her first internship or entry-level job in a major American city. Wealth itself refers to the ownership of capital. This includes real estate, stock and bonds, oil fields, and any other investment that can potentially generate a profit for investors. Those with wealth by definition have a much more stable financial ground to stand on than those without it; a standing parents readily pass on to their children in terms of private schooling, tutoring, music camps, special athletic training, and unique vacation experiences among other things, given the opportunity.

One issue that has caught the public's attention in the first part of the 21st century is the high concentration of wealth among a small minority of the population. According to the Congressional Budget Office (CBO), in 2013 the top 10 percent of households in the United States held 76 percent of the nation's family wealth, whereas the bottom 50 percent of households possessed just 1 percent (Congressional Budget Office, 2016). Wealth concentration has steadily increased over the past 30 years with households near the top at the 90th percentile increasing their wealth on average by 54 percent – families in the top 10 percent of the wealth distribution had on average $4 million dollars in assets. Meanwhile, households at the lower end of the scale – the 25th percentile – saw their wealth decrease on average by 6 percent during the same time period. Using the term wealth in this last case is a bit of misnomer since the average family at or below the 25th percentile in 2013 had a net debt of $13,000.

The CBO's data is helpful but, by its own admission, incomplete. It is based on the Survey of Consumer Finances carried out by the Federal Reserve Board only every three years. This data is not itself tied to more specific demographic data that could provide greater detail such as where on the map wealth concentrates and among which groups. By its nature, wealth is hard to track. One might also add that in a nominally meritocratic, individualistic culture such as our own, wealth can sometimes bump into dominant assumptions regarding people's self-worth. The fact that wealth can accumulate and be passed onto the descendants of the wealthy, tends to undermine notions of "self-made" wealth. Though there exist amazing cases of people truly working their way out of poverty and into the upper strata of society – Oprah Winfrey, for example – they are extremely rare. Upward social mobility of that nature, though much more likely in the United States between 1948 and 1972, is the great exception rather than the rule (Gordon, 2017).

Acknowledging the invisible hand of inheritance in one's own success story is thus a bit of a buzz kill and, moreover, can be a little embarrassing. Witness in 2019 the scandal surrounding a number of wealthy US families who had hired "an agent" to help get their children admitted into an Ivy League school by forging documents, lying about athletic accomplishments, and even feigning learning disabilities (Jaschik, 2019). The disingenuousness of these actions is appalling, but they also call attention to any number of other perfectly legal strategies wealthy parents can take to improve their offspring's chances in life. These include living in a good neighborhood with a well-funded school system, paying for test-taking courses specific to standardized college entry exams, and if they are fortunate enough, taking advantage of legacy admissions in the Ivy Leagues and other private institutions which give preferences to the children of those schools' alumni.

Education

Wealth thus obviously plays an important part in determining our place in the social hierarchy. Why, however, are so many rich people obsessed with where their kids go to school? Not coincidentally, education is also another key measure of SES used by sociologists interested in social stratification. Where, what, and how long you studied in school, as it happens, are all excellent "predictors" of your success in life, including how much money you earn and what kind of person, socioeconomically speaking, you are most likely to marry. To put it in Emile Durkheim's words, education is also an excellent "social fact." That is, as data goes, it is readily available from innumerable sources, and any sociologist worthy of the name will know to ask their research subjects what is their level of "educational attainment." What's more, in any given survey, people are much more likely to volunteer the level of education they have attained than their annual household income. Similar to income, however, the same level of education may have a different value depending on where you got your diploma, hence the disproportionate cultural value given to an Ivy League education.

From what I have heard, the odds of receiving quality instruction from Harvard, Yale, Princeton, and the like are pretty good. However, the exorbitant amount of resources dedicated to getting teenagers into institutions that accept less than 5 percent of applicants suggests a battle royale within some sectors of society for the added social status conferred by an Ivy League education beyond what is being learned in the classroom. A more generous assessment of this situation would acknowledge that the linkages to elite social networks afforded by these institutions may also to a large degree justify the price of admission. Either way, these particular angles on prestige and social capital suggest that education has a much greater cultural meaning in society beyond the mere accumulation of valued skills in the job market or even satisfying one's intellectual curiosity through learning.

Clearly, education is also about signaling status, and thus once again we are confronted with our loaded cultural assumptions about merit and individualism. Given the pricey market of conspicuous educational attainment, to what degree do our academic achievements reflect our hard work, determination, and stick-with-it-ness, versus simply reflecting the level of privilege or lack thereof we were born into? As is often the case for sociological questions like this, the answer is a little bit of both. Regarding willpower and determination, you have probably heard an educator or parent talk about the importance of the "growth mindset," an emphasis on intelligence as a learning process as opposed to an innate ability. This is, of course, a healthy strategy for approaching academic challenges as it underscores the importance of effort in attaining one's goals. However, handy psychological tricks such as this do little to overcome the enormous obstacles in the way of reducing the substantial achievement gaps in education among children of different SES backgrounds.

One tactic taken by sociologists who study inequities in education is to consider the relative impact schools versus family life have on children's academic performance. Beginning with schools, one of the major concerns has been the difference in funding received by public schools in relatively wealthy versus relatively poor neighborhoods. Seeing this as a key driving force behind unequal outcomes in education, places such as my own state of Vermont have made commendable efforts to redistribute property tax revenue equally across the state, thus assuring public schools have similar levels of financial resources, regardless of their place on the map. Policies such as these really do make a difference, especially with regard to providing well-maintained facilities and staffing in districts with a lower tax base. Sociological research in this area, however, is mixed. In 1964, James S. Coleman, who we mentioned in an earlier chapter on multiplexity, was charged by Congress with leading a national survey to understand the impact of unequal access to resources among public schools in the context of racial segregation. His rather astonishing conclusion from this research is summarized as follows in the Coleman Report:

> Taking all these results together, one implication stands out above all: [S]chools bring little influence to bear on a child's achievement that is independent of [the child's] background and general social context... [This] means that the inequalities imposed on children by their home, neighborhood, and peer environment are carried along to become the inequalities with which they confront adult life at the end of school.
>
> (Coleman, 1968, p. 325)

Over 50 years later, perhaps the most remarkable thing about the Coleman Report is that, despite critiques of its outdated methodology, its central finding has now been tested by other researchers many times and remains valid today. The inequalities children face outside of school, especially as indicated by their families' economic status, far outweigh the level of resources provided in the classroom when it comes to predicting how they do in school.

This is not to say that schools do not matter – the quality of individual teachers, in particular, appears to have a significant impact on how well a child learns. However, gaps in educational attainment between students of different SES backgrounds have changed little since the publication of the Coleman Report and tend to underscore the notion that one's education is still more a reflection of pre-existing inequalities than it is an objective equalizer of opportunities in a diverse society.

To be clear – stay in school; there is plenty to learn there, and so long as you are not overly reliant on student loans to pay your way, the long-term financial returns from a college education are still mainly worth it. Therein, however, lies the rub – one's ability to benefit from a quality education these days is highly dependent on the economic circumstances into which one is born including, significantly, the ability to avoid onerous debt.

In her book, *Unequal Childhoods: Class, Race, and Family Life*, sociologist Annette Lareau gets inside the family dynamics that strongly determine the academic trajectory of children from an early age (Lareau, 2011). Inspired in part by Bourdieu's notion of the habitus described above, Lareau's ethnographic research calls attention to the different parenting styles evident in working-class versus middle- and upper-middle-class families. In working-class households, Lareau sees a tendency for parents to fall back on "natural growth" strategies which give children a relatively unstructured schedule, allowing them to decide for themselves what to do with their free time. In many ways, this sets the stage for a childhood grounded in strong engagement with family and friendships with children who live nearby.

By contrast, parents from higher income households are more likely to utilize "concerted cultivation" strategies which, as the name implies, involves a deliberate effort to structure their children's time in ways that will improve their odds in school and life more broadly. Along with the intense scheduling of extracurricular activities, Lareau captures the very specific styles of interactions between highly educated parents and their children. These include, the proper use of language, maintaining eye contact, assertiveness, and confidently interacting with adults in positions of authority; all things which, for the children of relative privilege, will serve as a lifelong set of advantageous social skills.

Occupational status

There is one last measure of SES that also tells us something about how people in modern life are sorted and selected into groups of people similar to themselves – the essence of homophily. Along with having one of the more iterative names in sociology, Otis Dudley Duncan gained notoriety in the 1960s for his research on occupational status (Blau & Duncan, 1967). Duncan and his colleagues discovered in their survey work a pattern that has proven consistent over time – when given a list of common occupations, respondents tended to rank them in a predictable order with physicians, dentists, and lawyers near the top, electricians, mail carriers, and bank tellers near the middle, and janitors, prostitutes, and panhandlers near the bottom. Though these surveyed ranks of occupations tend to track with levels of individual income, there is not always a direct correlation. Occupational status is thus itself often used as a separate measure of SES that, unlike income, can reflect common biases and cultural assumptions of status across different regions of the country. This brings us back to the question of methodology: Given the incomplete views of SES provided by any single measure of income, wealth, or education, occupational status allows us to triangulate with these other variables providing a reasonable approximation of what we mean by an individual's standing in a world stratified by social class.

There is a risk here, however, which we would prefer to avoid when thinking of the social world in terms of individual measures of SES; namely, *methodological individualism*. This refers to the tendency when dealing with social science data to think of people as primarily the list of individual-level characteristics we have collected on them (Udehn, 2002). Along with being an extraordinary act of reductionism vis-à-vis the human experience, describing social life in this way belies a central tenet of sociology – understanding social context. How do people connect (or not) to one another? Do they act cooperatively or in fear of others not like themselves? What effect do high levels of socioeconomic inequality have on the ability to generate mutual trust in a diverse society? Under what circumstances do people feel empowered to act collectively to bring about social change? These are the kinds of questions not easily answered with only individual-level data at our disposal. From this perspective, social stratification is really more the stage and scaffolding upon which the actual drama of social life takes place.

Capturing interdependence through network analysis

Often coming from a more quantitative perspective, social network researchers in sociology have made important advances in trying to get an empirical handle on the interdependent nature of life among other humans. Mark Granovetter's 1973 article, "The Strength of Weak Ties," for example, is one of the most highly cited articles in the discipline of sociology (Granovetter, 1973). In it, he describes the different kinds of resources obtained from different kinds of relationships. From our good friends and trusted family members – "strong ties" in Granovetter's parlance – we find reliable sources of emotional and even, when necessary, financial support. These are essential ties that can inform our most important decisions and are indispensable when we find ourselves in times of trouble. Strong ties, however, are not a particularly good source of new information.

For instance, are you looking for a new job, or trying to find quality childcare at a reasonable price in a new location? In these cases, Granovetter suggests you consider talking with your acquaintances, the people you do not know so well. Or, better yet, the people you do not yet know but will encounter, for example, at a social event hosted by the friend of a friend. But, why would these "weak ties" be a better source of new information than our strong ties to the people we know best? The answer to this question brings us back to the underlying theme in this chapter – homophily. That is, the people to whom we are closest tend to know the same things we do, in part because we spend more time with them, but also because our social circumstances tend to be very similar. In contrast, step out a few degrees from your core social circle and a universe of novel information and opportunity opens up to you. Yes, as the ancient Greeks said, to have a meaningful life one must "know thyself." However, one of the great lessons of contemporary network theory is (with apologies to the Oracle of Delphi) we would do well to "know others not like thyself," as well.

Research on socioecological resilience, for example, finds that bringing together people who would not normally interact – often because of a history of mistrust and power differentials between affected parties – can be an effective way to begin to work out natural resource conflicts (Paolisso et al., 2019). When interaction is facilitated among local communities, government officials, industry, and academics, common interests are more easily perceived, and a way forward in the face of collective threats such as forest fires, flooding, and weather volatility is possible. Informal networks resulting from these interactions have been shown to be better conduits of information than formal organizations (Prell, Reed, Racin, & Hubacek, 2010). Moreover, information and opinions passed through informal networks tend to have a stronger sway on stakeholders' views about environmental issues than those conveyed through formal organizational channels (Saunders, 2007).

This is not to say that participants in diverse collaborative projects always get along. One study on forest management in British Columbia, for example, found that greater interaction among stakeholders was not associated with a higher opinion of forest industry management among environmental activists (Stoddart & Tindall, 2010). Just because you interact with someone, does not mean you will like them. That said, social interaction is a fruitful source of meaning generation, grounded in direct experience. Storytelling and face-to-face conversations are excellent generators of empathy, trust, and goodwill, and almost always better than the alternative – i.e., not communicating at all (Lejano, Ingram, & Ingram, 2013).

We could stop here and see homophily as primarily a reminder to seek greater diversity in our personal relations with others; a way of accumulating social capital and cultural insight for our individual and collective advantage. This is often what people mean by the term "networking," i.e., getting the information and resources we need through the instrumental deployment of social ties outside our immediate social milieu. However, in the face of structural inequity, network solutions do not often suffice, and the collective impact of homophilic ties in a class-stratified society is cause for alarm. Individual preferences for people similar to us in education and wealth can accumulate in ways that lead to severe levels of neglect and even social isolation among certain groups in society. In the next chapter, we will see how our preferences for people like ourselves are strongly shaped by the correlates of inequality, including where we live, the color of our skin, our level of education, and even exposure to environmental hazards.

Glossary

Educational attainment One of the four components of socioeconomic status representing the sum of one's educational accomplishments.

Habitus The physical embodiment of cultural capital; an idea coined by Pierre Bourdieu.

Household income While not the sole determinant of socioeconomic status, it is the most widely understood, representing the dollar amount earned for paid labor collectively among individuals in a household.

Methodological individualism Viewing an individual as a list of their demographic characteristics and social status rather than an individual with real experiences that transcends a list of traits and qualifications.

Occupational status One of the four determinants of SES, occupational status reflects the perceived ranking of occupations in society, from least to most esteemed.

Socioeconomic status (SES) One's ranking among a collection of socially-constructed and hierarchical categories reflecting income, wealth, education, and occupation.

Stratification Social layering indicative of inequality and hierarchy in society. Historically and currently there exist a number of different systems of stratification, including slavery, caste, and feudalism.

Strong ties Our close social connections with other whom we rely on for emotional and psychological support.

Weak ties Somewhat distant social relationships that can be advantageous, especially as regards access to new, helpful information and opportunities not likely accessible through one's established close ties.

Wealth One of the four determinants of SES representing the value of all of one's financial assets.

Questions

1   How *would* you describe your habitus? Do you think people who have never met you before could place you in the socioeconomic scheme of things with just a brief interaction? Have you ever found yourself doing this with other people you have just met? Consciously or unconsciously, do you think this has played a role in the way you select your friends and social circles? Do you ever work against the homophilic tendency to prefer the company of people similar to yourself?

2   What system of stratification do we currently live in? Officially, President Lincoln declared the abolition of slavery in the United States in 1963. Since that time, however, there has been some debate as to whether we live primarily in a class-based or caste-based system of stratification. On the one hand, we are not as workers bound to a particular employer or master, for that matter. On the other hand, there is evidence for ongoing discrimination and significant differences in outcomes and opportunities on the basis of race. What do you think?

3   Consider the four primary ways we gauge social class status in sociology –
income, education, occupation, and wealth. Which of these do you think
is the most reliable measure of what we mean by social class and why?
Which do you think is probably the least reliable and why?

References

Bales, K. (2012). *Disposable people: New slavery in the global economy, updated with a new
preface*: University of California Press.
Blau, P. M., & Duncan, O. D. (1967). *The American occupational structure*: Wiley.
Boehm, C., & Boehm, C. (2009). *Hierarchy in the forest: The evolution of egalitarian
behavior*: Harvard University Press.
Bottero, W. (2007). Social inequality and interaction. *Sociology Compass, 1*(2), 814–831.
Bourdieu, P. (1977). *Outline of a theory of practice* (Vol. 16): Cambridge University
Press.
Coleman, J. S. (1968). Equality of educational opportunity. *Integrated Education, 6*(5),
19–28.
Congressional Budget Office. (2016). *Trends in Family Wealth, 1989 to 2013*. Re-
trieved from https://www.cbo.gov/publication/51846. Accessed September 16,
2021.
Engerman, S. L. (1999). *Terms of labor: Slavery, serfdom, and free labor*: Stanford Uni-
versity Press.
Fry, R. (2014). New census data show more Americans are tying the knot, but mostly
it's the college-educated. *Pew Research Center*, February, 6.
Gordon, R. J. (2017). *The rise and fall of American growth: The US standard of living since
the civil war* (Vol. 70): Princeton University Press.
Granovetter, M. S. (1973). The strength of weak ties. *American Journal of Sociology,
78*(6), 1360–1380.
Huber, G. A., & Malhotra, N. (2017). Political homophily in social relationships:
Evidence from online dating behavior. *The Journal of Politics, 79*(1), 269–283.
Jaschik, S. (2019). Massive admissions scandal. *Inside Higher Ed*. March 13. Re-
trieved from https://www.insidehighered.com/admissions/article/2019/03/13/
dozens-indicted-alleged-massive-case-admissions-fraud. Accessed September 16,
2021.
Lareau, A. (2011). *Unequal childhoods: Class, race, and family life*: University of Cali-
fornia Press.
Lejano, R., Ingram, M., & Ingram, H. (2013). *The power of narrative in environmental
networks*: MIT Press.
Lenski, G. E. (2013). *Power and privilege: A theory of social stratification*: UNC Press
Books.
Livingston, G., & Brown, A. (2017). Intermarriage in the US 50 years after Loving
v. Virginia. *Pew Research Center Report*, 1–35.
McPherson, M., Smith-Lovin, L., & Cook, J. M. (2001). Birds of a feather: Homo-
phily in social networks. *Annual Review of Sociology, 27*(1), 415–444.
Musick, K., Brand, J. E., & Davis, D. (2012). Variation in the relationship between
education and marriage: Marriage market mismatch? *Journal of Marriage and the
Family, 74*(1), 53–69. doi:10.1111/j.1741–3737.2011.00879.x

Paolisso, M., Prell, C., Johnson, K. J., Needelman, B., Khan, I. M., & Hubacek, K. (2019). Enhancing socio-ecological resilience in coastal regions through collaborative science, knowledge exchange and social networks: A case study of the Deal Island Peninsula, USA. *Socio-Ecological Practice Research*, *1*(2), 109–123.

Prell, C., Reed, M., Racin, L., & Hubacek, K. (2010). Competing structure, competing views: The role of formal and informal social structures in shaping stakeholder perceptions. *Ecology and Society*, 15(4), 34.

Sahlins, M. (1998). The original affluent society. Gowdy, John, ed. *Limited Wants, Unlimited Means: A Reader on Hunter-Gatherer Economics and the Environment* (pp. 5–41): Island Press.

Saunders, C. (2007). Using social network analysis to explore social movements: A relational approach. *Social Movement Studies*, *6*(3), 227–243.

Stoddart, M. C., & Tindall, D. (2010). 'We've also become quite good friends': Environmentalists, social networks and social comparison in British Columbia, Canada. *Social Movement Studies*, *9*(3), 253–271.

Udehn, L. (2002). The changing face of methodological individualism. *Annual Review of Sociology*, *28*(1), 479–507.

Veblen, T. (2017). *The theory of the leisure class*: Routledge.

7 Social and environmental justice in a diverse society

Sociological writing – including here journalists versed in the sociological literature – has a long history of providing qualitative, first-person accounts of people whose lives are not typically represented in popular culture and media. In the book *Nickel and Dimed* (2010), Barbara Ehrenreich does her best to live off the wages paid to her in a series of working-class "pink collar" (mainly female-occupied) jobs in southern Florida. With a PhD in biology and a successful writing career, Ehrenreich by her own admission is not playing fair: She allows herself a "start-up" allowance of $500 while conducting this research, enough to cover deposit for rent. Nonetheless, in her various stints which include waiting tables, toiling for a domestic cleaning service, and working at Walmart she finds it essentially impossible to make ends meet on her paltry earnings.

How, she asks, can policymakers expect the "working poor" to simply labor their way out of poverty, especially at a time when the government has systematically decreased assistance for those most in need? Written in the late-1990s, *Nickel and Dimed* shows the opportunity structure for upward mobility to be severely undermined, especially when compared to the post-World War II era of high economic growth. On a hopeful note, Ehrenreich did find a strong sense of solidarity and friendship among her coworkers despite the fact that many of them were temporarily housed in motels or living out of their cars.

More recently, sociologist Matthew Desmond has looked at the plight of the poor in the wake of the 2008 financial crisis. Over a period of several years, Desmond lived with and took down an account of the lives of eight families in Milwaukee, Wisconsin who did what they could to pay rent and not get evicted from their homes. In many of these cases, the quality of the rented living spaces was by many standards uninhabitable. However, given the precariousness of their existence – too many 911 calls, noisy children, and missed rental payments were all grounds for eviction – tenants were unlikely to complain to landlords about living conditions. Eviction – which by Desmond's account is typically a quick, brutish act of domestic purging by law enforcement – meant homelessness for most of these families.

DOI: 10.4324/9781003110668-7

His book, *Evicted* (Desmond, 2016), is not an easy read. That is, it is well-written, winning the Pulitzer Prize for non-fiction in 2017, but it takes the readers to decrepit places and humiliating experiences that many readers have never personally encountered. Beyond the theme of exploitative housing – eviction lawyers make a good living; the proprietor of a mobile-home park in the book earns over $400,000 a year – Desmond shows us the devastating effects crack cocaine, opioid addiction, and alcoholism have had on families at this level of the social hierarchy.

The Bogardus scale: a century of social distance measurements

Ethnographic research such as that by Ehrenreich and Desmond can give us a visceral sense for what early 20th-century sociologist Emory Bogardus referred to as the *social distance* between us and the people whom, though we live within the same society, we may rarely encounter in our everyday lives (Bogardus, 1926). The "Bogardus scale," first formulated in the 1920s, also captured attitudes of prejudice as it surveyed people about their personal preferences for different groups in society. After receiving a list of various US race and ethnic groups, the survey respondents were asked to gauge the level of social proximity they would tolerate with a member of each group.

To give but one example, Would you accept a Bulgarian as a family member by marriage (1 point)? Would you accept a Bulgarian as a close friend (2 points)? How about as a neighbor (3 points); or perhaps a co-worker (4 points); all the way down to, would you prohibit Bulgarians from entering the country (7 points)? The smaller the amount of points allotted, the shorter the social distance between the respondent and the ethnic/racial group in question. In his 1946 study immediately after World War II, the categories "Americans" and "Canadians" were at the top of the list of groups ranked closest to furthest by social distance in the United States. At the bottom of this list were "Negroes" and the "Japanese" (Bogardus, 1967).

Up to now, we have focused primarily on social class as a determining factor in the social sorting that takes place among groups in American society. The Bogardus scale, however, makes clear that race and ethnicity have also long played a key role in how we decide with whom we socialize. In 1946, for example, Americans had come out of a global conflict that required much sacrifice – including hundreds of thousands of lives – throughout the population. It thus might be understandable that the Japanese would be at the bottom of this ranking since Japan was one of our major adversaries during World War II. However, Germany was also a major adversary during that war. How can we explain the category of "German" being ranked only 11th in a list of 36? Moreover, African American troops fought heroically during the war in all major branches of the armed forces. How is it they were ranked second to last using this same measure of social distance? Beyond our levels of education, income, and preferences for food, music, and other consumer items, the Bogardus scale reveals that bias for and against certain race and ethnic groups is a significant source of social sorting in American life.

The somewhat good news regarding the Bogardus scale is that, as it has been replicated over the years the average social distance between groups has steadily decreased (from 2.14 in 1926, to 1.45 in 2001). This would tend to suggest that we live in a more tolerant society than what existed almost a century ago. Before we celebrate our nation's acceptance of diverse people, there are a few flies in the ointment which should give us pause before embracing this finding as an inevitable trend. First, attitudes, especially related to race and ethnicity can be tricky to measure. *Social desirability bias* refers to the tendency of survey participants to respond in ways they believe will be perceived positively by researchers (Krumpal, 2013). Arguably, social norms in the way we talk about race and ethnicity in the United States have shifted dramatically over the last century and overtly proclaiming preferences for specific race or ethnic groups is broadly frowned upon (even if there are some indications this norm is shifting in the opposite direction). Swallowing whole the notion that we are becoming more tolerant as a nation based on an attitudinal survey is not recommended.

Second, though in the most recent iteration of the Bogardus scale African Americans are now equal to German Americans in their relatively low levels of perceived social distance, in the post-9/11 era of the early 21st century, other groups are now relegated to the bottom of the list, notably "Muslims" and "Arabs." We can only assume that an even more recent iteration of the Bogardus scale taken after the 2016 election would reveal large average perceived social distances between respondents and populations of migrants and refugees who have been depicted by some political leaders as an existential threat to our identity as a nation. Using this particular survey instrument, it can be hard to disentangle an actual measure of social distance from the political mood of the times, though one might also argue that these two elements of modern life are perpetually intertwined.

Finally, there is a good deal of evidence that we are as segregated as a nation by race as we were at the end of the Civil Rights era in the 1960s. Here, we shift from the attitudinal perception of social distance to the physical separation of people of different backgrounds in the contemporary landscapes of housing, education, and occupations; from prejudicial beliefs to discriminatory actions. Housing, education, democracy, and work were the key areas addressed by the Civil Rights Movement in the 1950s and 1960s, and the resultant legislation passed at that time. The Supreme Court decision of *Brown versus the Board of Education*, the Civil Rights Act of 1964, the Voting Rights Act of 1964, and the Fair Housing Act of 1968 mark major turning points in terms of the legal recourse available to address racial discrimination in the United States.

Focusing on one area in particular we see that, prior to the Civil Rights era, racial segregation in housing was not only tolerated, but *de jure* – i.e., officially sanctioned by law – throughout most of the country. This may come as a surprise if you were led to believe that the main historical divide of racial discrimination was along the Mason-Dixon Line between the North and South. In fact, at least two instruments of legal segregation were present all over the country in the pre-Civil Rights era – the so-called "redlining" and housing covenants.

The structure of segregation in American society

To understand *redlining*, it is important to understand its origins in good intentions. Back in the 1930s, the United States and the world were in the depths of the Great Depression. Unemployment was around 20 percent, economic growth had withered, and there appeared to be no end in sight to persistent poverty that prevented people from buying the things they needed to survive. The country was in dire need of economic stimulus. It was in this milieu of desperation that the administration of Franklin D. Roosevelt established in 1934 the Federal Housing Administration, a program so ingenious and successful, it still exists today. Instead of simply giving money away, the US government agreed to guarantee bank loans for the purchase of a home. At the time, more than half of American families rented their place of residence, thus assuring a growth market in real estate should this demographic segment take the bait. Moreover, economists knew that buying a house entails much more than just ownership of property – rooms must be painted and decorated, furniture must be purchased, appliances must be installed, and as any experienced homeowner can tell you, maintenance is perpetual.

The modern American Dream was thus born and subsidized through bank loans guaranteed by the US government. The catch, however, was that the dream would not be available to all equally. Banks in major cities across the country began to include normative prejudice against African Americans in their calculations of who could and could not get access to a federally subsidized loan. City maps were drawn up showing more and less desirable housing stock for loan investment. Consistently, the least desirable neighborhoods – often outlined in red – were also where the greatest concentration of Blacks resided in the inner city (Badger, 2017).

By the end of World War II, segregation in major American cities was actually increasing as returning GIs and their families with easy access to loans left central urban areas in droves through a process sometimes referred to as "white flight" due to the racial imbalance evident among the people populating the newly built suburbs. It was at this point that even the term "urban" began to take on the meaning of ethnic and racial minority-concentrated inner-city areas. Meanwhile, as housing values in the suburbs tended to increase over subsequent decades, home prices in urban areas tended to stagnate, thus effectively keeping an important source of wealth generation in American life out of reach for many African American and other non-white homeowners (Rothstein, 2017).

Another form of *de jure* segregation at this time was what is known as housing covenants. In many communities around the country – prior to the passage of the Fair Housing Act in 1968 – it was common practice to include in the purchase agreement or title of a house, specifications about what kind of people, based on race or ethnicity, could live there (Rothstein, 2017). In 1952, my father, a World War II veteran and US-born son of Mexican immigrants, was looking for a property upon which to build a house in Phoenix with my mother, also a second-generation Mexican American. After finding a property in an area they liked and could afford, they were frustrated many times to discover that the local community had collectively agreed to prohibit non-whites from living there.

Even after finally purchasing a parcel of land from the sellers who would become our longtime neighbors – the Brevermans – my mom discovered there had been an unsuccessful effort among a local group to prevent my parents from closing on the property. There is a somewhat hopeful coda to all this. By the 1970s when I was growing up and began my afternoon paper delivery route in the neighborhood, I got to know many of these people who seemed like kind retirees who enjoyed small talk and even thought to tip me once in a while. As far as I could tell, 25 years after my parents moved in, racial animosity in the neighborhood had dissipated considerably.

Not all families, however, were as fortunate as mine, and among communities of color in the United States, African Americans have experienced especially severe levels of residential segregation, something sociologists Doug Massey and Nancy Denton have referred to as *hypersegregation* (Massey & Denton, 1993). One of the key measures of hypersegregation is the index of dissimilarity (ID), measured on a scale of 0 to 1. The essential question here is, do group proportions scale down? For example, let's say there exists a hypothetical city wherein 80 percent of the population is white and 20 percent is Black. A score of 0 on the ID would indicate that these percentages scale all the way down. That is, in every neighborhood, on every block of this city we find the same percentage breakdown – 80 percent white, 20 percent Black. This would describe a perfectly integrated city where you would likely have social interactions with someone from both groups on a daily basis. A score of 1, by contrast, would indicate a perfectly segregated city – an apartheid, really – where all Blacks live in one part of the city and all whites live in another part, assuring on average very little interaction between the two groups, at all.

In reality, neither extreme of 0 or 1 on the ID exists, but a large range between these two scores exists among American cities. Massey and Denton and others have argued that a score for the ID that exceeds 0.6, is indicative of hypersegregation. Going by that standard, many post-industrial cities of the Northeast and upper Midwest qualify as hypersegregated for African Americans, including Chicago, Cleveland, Detroit, Milwaukee, Newark, and Philadelphia. But, so what? This is a free country, the Fair Housing Act was passed over half a century ago, and well, homophily – people like to be with people like themselves; ergo, segregation is a natural outcome of living in a free market society.

To some degree this is true, when left to their personal preferences, there is a tendency for economically dominant groups to choose to live in neighborhoods populated with people like themselves – a practice which disproportionately excludes underprivileged communities of color, especially African Americans. These are choices "freely" taken by individuals with sufficient economic power to make them. They, however, are not "natural" if we consider the legacy of advantages or disadvantages into which different groups are born.

The unnatural practices of racial discrimination

Moreover, when an entire class of people is excluded from opportunities on the basis of "common sense" assumptions about their worthiness, the term homophily is no longer sufficient. We have now moved beyond a casual preference for people with similar tastes and proclivities into the realm of stereotypes, prejudice, and racism. Unfortunately, despite the passage of Civil Rights legislation over 50 years ago, evidence of racial prejudice and discrimination persists. An effective way of testing this assertion in the social sciences is to use "matched-pair" methodology. In this kind of study, a researcher will hire a set of actors who are matched on every characteristic – gender, occupational status, education, etc. – except race; e.g., one set of actors are Black, the other, white.

One of the more remarkable illustrations of this was a study conducted by sociologist Devah Pager who, in the early 2000s, hired Black and white match-paired actors to look for employment in Milwaukee (Pager, 2003). As an additional twist she included another variable characteristic – half of the actors informed their potential employer they had committed a felony, while the other half reported a clean record. Given the history of discrimination in the United States, it was unsettling, if not surprising, to find out that white job applicants were twice as likely to get a call back for an interview than their matched Black counterparts. More disturbingly, Pager also found that whites with a reported criminal record were still more likely to get a call back than Blacks without a criminal record. That is, in that particular market for jobs, the stigma for being Black was greater than the stigma of having committed a felony.

Pager's research is consistent with contemporary findings from other matched-pair research in the housing market where African American couples have been disproportionately "steered" by real estate agents to predominantly African American neighborhoods (Pearce, 1979; Turner, 2013). Her work is also consistent with other work in hiring practices where job applications are sent out to employers with the names of fictitious applicants matched for qualifications; the one difference being half the applicants have African American-sounding names and the other half does not (Bertrand & Mullainathan, 2004). The result – a significantly higher proportion of callbacks for people from the second group.

It is important to understand that this current wave of research on racial discrimination in the job and housing markets takes place after the passage of federal laws established that make these types of actions unconstitutional. That is, we have moved from an era of *de jure* to one of *de facto* discrimination wherein discriminatory behavior, though not ordained by law, is carried out in everyday practice, nonetheless. At a time when overt expressions of racial preferences are frowned upon, sociological research such as that carried out by Pager, Massey, Denton, and others, is often the primary evidence we have of the broader pattern of discriminatory actions that persists today.

Beyond the blow to one's self-confidence of being systematically left out of opportunities in the job and housing markets, the material impact of not being able to improve your occupational status or increase the equity value of your home can have lifelong if not generational significance for families on the wrong side of the racial divide. Persistent discrimination and segregation can lead to the perpetuation of what sociologist William Julius Wilson calls the American *underclass* (Wilson, 2012). This refers to large groups of people segregated by class and race in American society who live in impoverished neighborhoods with underfunded schools and social services, few economic opportunities, high rates of crime and violence, and low quality housing stock. This portion of the US population has long lived outside the view of the majority population.

The past decade, however, has seen a resurgence of activism within this sector of society. The Black Lives Matter Movement has done much to call attention to racial inequality and the violence inflicted on young Black men in particular whose portion of the US prison population (40 percent) is about three times their portion of the general population. The widespread use of social media calling attention to racial injustice and the publication of Michele Alexander's book, *The New Jim Crow* (2012) have both strengthened this movement which seeks to bring the lived experience of racial exclusion back into the consciousness of the majority population and which we will return to in the final chapter of this book.

The place of race in disposing our waste

A final area crucial for our understanding of the material outcomes of social sorting is environmental justice (EJ). Who is and who is not most directly exposed to environmental hazard in towns and cities and the world more broadly? EJ activism goes back to a time not long after the end of the Civil Rights era. In fact, many of the people involved in the early protests for EJ cut their teeth during the marches, sit-ins, and other acts of civil disobedience that took place in the 1950s and 60s. The first protests associated with the term "environmental justice" took place in Warren County, North Carolina where, in 1978, industrial polluters deliberately dumped 31,000 gallons of oil laced with highly toxic polychlorinated biphenyls (PCBs) alongside 240 miles of regional highway (Bullard, 2018).

Four years later, the state government decided to place 120 million pounds of soil contaminated from this illegal dumping in a landfill in Warren County. At the time, 69 percent of the population in this part of North Carolina was African American. Protesters gained national attention when, using the tactics of civil disobedience, they blocked the roadway, lying down in front of dump trucks attempting to transport toxic dirt into their part of the state. Subsequent research showed that there were places better suited geologically for a toxic dump than Warren County. However, the racial and socioeconomic makeup of the region suggested that other demographic considerations – i.e., race and socioeconomic status – were more important.

Broader quantitative studies in EJ have found there to be a consistent pattern across the United States – the populations near and around environmentally hazard sites such as waste incineration plants, toxic dumps, and chemical plants tend to disproportionately be made up of African Americans and Latinos, as well as lower income groups, more generally. The underlying theme here is that across rural and urban landscapes, environmental "bads" – specifically, polluted air and water, and the toxic waste resulting from industrial production – are unequally distributed within the population. Beyond the day-to-day nuisance of exposure to foul smelling air, and the perpetual flow of waste removal vehicles passing through one's neighborhood, residents in these areas often have higher rates of diseases such as childhood leukemia and other kinds of cancer associated with regular exposure to noxious chemicals.

Both Blacks and Latinos living in segregated urban areas face higher risks of cancer caused by air pollution than do members of the majority population. In this last regard, a recent US study published in the *Proceedings of the National Academy of Sciences* (Tessum et al., 2019) is quite specific in its accounting. Integrating data from the EPA's National Emissions Inventory, the Bureau of Labor Statistics, and the Bureau of Economic Analysis, researchers found that, on average, white Americans experience about 17 percent less air pollution than they produce through the consumption of goods, services, energy, and transportation. Meanwhile, Blacks and Hispanics are exposed to 56 and 63 percent more air pollution, respectively, than what they produce through their consumption.

Since Bullard's work in the 1980s, the dominant vein of EJ research within sociology has been empirical and descriptive, calling attention to a form of geographic inequality other measures of socioeconomic status – e.g., income, wealth, education, occupation – were simply not capturing. In recent years, drawing on the momentum of the Black Lives Matter movement, struggles over native sovereignty and immigrant rights, a group of sociologists has begun to take a more critical stance on the nexus of environment and society. This perspective begins with a critique of the logical corollary of the fact-finding approach to EJ – *distributive justice*. That is, if it is simply a matter of some groups being more highly impacted by environmental hazards than others, the solution is to redistribute risks more equitably throughout social geographic space. In this way, everyone, rich or poor, Black or white, would be equally exposed to air and water pollution, for example.

Moving beyond distributive justice

Through Critical Environmental Justice (CEJ) Studies, sociologist David Pellow and others have argued that the EJ goal of distributive justice is not enough. In fact, nothing short of a rejection of the status quo is required to bring about needed changes (Pellow, 2017; Pellow & Brulle, 2005; Sbicca & Myers, 2017). To do this, Pellow spells out four pillars indicating a radical refashioning of EJ research within sociology.

First, *intersectionality* points to the way multiple categories of difference – beyond race, gender, and social class, this includes "more than human" actors such as plants, animals, and the natural environment – are both subject to oppression by dominant actors and potential agents of socioecological change. In addition to the emancipatory influence of Marxist thought, this pillar is informed by ecological feminism, queer theory, disability studies, and critical animal studies. People of color, LGBTQ persons, immigrants, and the poor, among others, represent a diverse array of lived experience. From the CEJ perspective, however, they all likely share the common experience of mistreatment, exploitation, and socioeconomic exclusion from dominant groups who stand to benefit from the inferior status of others. It is this common set of grievances, however, that weaves together intersectionality, fosters solidarity, and makes possible a liberating response against domination.

The second pillar of CEJ is *scale*; an aspect essential for overcoming the "not-in-my-backyard" strain of EJ research that tends to neglect the broader structural origins of place-specific environmental hazards. Higher rates of respiratory illness and lung cancer in neighborhoods proximate to waste incinerator plants are an immediate concern, threatening the lives of individuals and families where they live. However, a broader perspective on the quantity of garbage produced in a society that throws away almost half of what it consumes allows us to consider the place of a particular environmental threat within a wider context of industrial production and consumption.

Scale through this perspective also refers to the passage of time. In the case of the Love Canal community in northern New York State, higher rates of childhood leukemia first brought to national attention by parent/activists in the late-1970s, were a result of noxious chemicals being buried in the ground a quarter century before by the Hooker Chemical Company. The legacy of oppression and environmental impact can stretch out for decades, centuries, and even millennia. The decision by European colonizers to enslave African people and forcibly remove indigenous people from their land represent legacies of violence and oppression that are still playing out today. Looking into the future, as mentioned in an earlier chapter, the nuclear waste resulting from the production of atomic energy will remain radioactive and a potential threat to living beings for tens of thousands of years.

Perhaps the most radical aspect of CEJ is its third pillar – *anti-statism*; i.e., a deep skepticism towards the role of government in bringing about socioecological change. This, as Pellow reminds us, is in contradiction to the strategies taken by more mainstream social movements. The Women's Suffrage Movement, for example, labored tirelessly to persuade those in power – i.e., white men – to grant half the adult population in the United States the right to vote in 1920. Similarly, the Civil Rights Movement appealed to dominant political institutions, especially the judicial and legislative branches of government, to attain its greatest victories in the mid-20th century. These strategies are also reflected in the efforts contemporary EJ activists who seek to gain the attention of those in power in the hope of having the economic and political power of government resources on their side. CEJ scholars, however, wonder if this deal with the devil – that is, powerful leaders who decide who gets access to resources and who does not – is really worth the effort.

On closer inspection, with respect to social change, national governments have a checkered past regarding their treatment of race and ethnic minorities, women, LGBQT people, and the natural world, and it could well be argued that the primary purpose of government is to manage and control – violently, when necessary – all within its dominion including human populations and natural resources. Given a long history of governments enforcing racist, sexist, and speciest ideologies on the living things within its territory, perhaps EJ activists could better focus their efforts trying to achieve more immediately tangible goals.

To be clear, CEJ scholars, on the whole, do not advocate getting rid of the state. As the women's rights, civil rights, and even environmental movements have shown, the state can play an important role in bringing about needed change. From the practical perspective of activists, however, those kinds of efforts take time, decades, usually. What really matters from a CEJ perspective is the here and now, actions that cannot wait for the approval of Congress, the Supreme Court, or even the mayor's office to sign off on. In this regard, Pellow advocates the practice of *direct democracy* wherein people and local communities deliberate, decide, and take action first. If the government is willing to follow their lead, that is fine, too.

The fourth and final pillar in the CEJ scheme is *indispensability*. In contrast to the ethic of expendability often evident in the treatment of marginalized groups by those in power, indispensability demands we recognize the inter-connectivity and interdependence of all life and matter on Earth. As opposed to sacrificing, eliminating, or assimilating groups into dominant cultures, the pillar of indispensability requires we respect the sovereignty of different origin people. In doing so, we also come to recognize that the diversity of human and more-than-human experience is essential for our survival.

My own work on environmental risk perception finds that race and ethnic minorities in the United States are not only more aware of environmental risks such as air and water pollution, and the harmful effects of climate change, but are also more willing to do something about it (Macias, 2016). Consistently, migrants from the global south, for example, are much more aware of the threat of water scarcity to human survival than are non-migrant residents in the global north (Carter, Silva, & Guzmán, 2013). Assimilation to dominant views, in this case, would not appear to be a particularly prudent strategy.

Perhaps unexpectedly, indispensability also applies to those in power. Pellow gives climate change as a primary example of this. The petroleum industry and climate deniers have convinced themselves that the perpetual contamination of the Earth's atmosphere through carbon emissions, though it may have more immediate impacts on vulnerable populations in the global south, will not affect them personally. Along with calling attention to the ways in which political and economic oppression has harmed the vulnerable, activists must also persuade those in power that progressive change is necessary for their own good.

As mentioned above, the more radical edge of the EJ movement has espoused an anti-statist stance (Pellow, 2017). Government, especially in recent decades, it is argued, has done more damage than good as it has opened the floodgates of industrial deregulation and done little to address the persistent unequal distribution of environmental "bads" in the United States and the world. Whether EJ anarchy is an effective means of addressing global climate change is unclear, but Pellow's analysis draws attention to the ways legacies of violence and exploitation can affect the ways people defined as racially different from the dominant population may have reason to doubt the state's willingness to work in their interest, including on questions of the environment.

The unnatural brunt of "natural disasters"

If we shift our gaze to the consequences of weather-related natural disasters, disproportionate impacts are also evident. Hurricane Katrina hit the coast of Louisiana on August 29, 2005 and is considered by many to be the worst storm to ever impact the continental United States, causing an estimated 1,200 deaths and over $100 billion in property damage. Prior to landfall, Katrina had stronger winds in the Gulf of Mexico, and the center of the storm actually made landfall further to the east along the coast near Biloxi, Mississippi. However, an ill-maintained system of levees in New Orleans and a shortsighted evacuation plan that did not account for severe transportation bottlenecks meant the effects of this "natural disaster" would be magnified by inadequate engineering and poor planning on the part of human beings.

As the days wore on, what also became evident was the disproportionate impact Katrina had on African Americans, a group long relegated to live in parts of New Orleans most vulnerable to flooding. From the families stranded on their rooftops waiting to be rescued, to the thousands of people temporarily warehoused in the Superdome before being bused off to other states, Blacks were visibly overrepresented relative to their portion of the population. The intersection of race with social class is also evident within the events surrounding Katrina (Elliott & Pais, 2006). The fact that it happened at the end of the month, for example, meant that the poor – many of whom anxiously await the first of the month for a paycheck or government assistance – could not afford a ride out of town before Katrina made landfall.

It is predicted that powerful storms such as Katrina will become more frequent as the oceans warm and weather in general becomes more volatile as a result of climate change. Other climate related changes such as severe drought and forest fires have also increased in frequency over recent decades and race and SES continue to shape the way people manage their way through these collective traumas, as well. When Hurricane Harvey hit Houston in 2017, undocumented immigrants – about a half million total live and labor in the greater Houston area – discovered that, as non-citizens, they would have no access to Federal Emergency Management Act (FEMA) funding (Florido, 2017).

More broadly, current research in sociology (Elliott & Howell, 2017) using national data finds that higher levels of FEMA funding in a disaster struck community are associated with *greater* social inequality after assistance has been distributed. The authors attribute this outcome to the focus on the recovery of property and wealth in this taxpayer-funded program versus the recovery of the social fabric. Property owners may regain some of the value of their investments, while the poor regain nothing and are often simply left to fend for themselves.

Environmental justice in the global context

At the global level, EJ has begun to have important implications for how we think of the ongoing environmental crisis. For example, in the 1970s, the US government passed a series of laws – e.g., the Clean Air Act, the Clean Water Act, and the Superfund Act – that went a long way in improving the natural environment and the quality of life for Americans who increasingly found themselves inhabiting a polluted environment. For the most part, air and water quality is better in the United States than it was 50 years ago, and the Superfund Act was impressive in the way it held industry accountable for paying for the cleanup of its own waste.

However, another important shift has taken place during this same period – *globalization*. Industrial production itself has moved overseas. Thus, the fact that most consumer goods we purchase in the United States are produced in countries with relatively little environmental regulation has permitted us as a nation a tremendous workaround. We can be smug and benefit from improved environmental standards since the 1970s, yet still pollute surreptitiously every time we buy goods made in another country where the output of waste from industry and transportation is relatively unregulated and where air and water quality has deteriorated over time.

The truth is, a sustainable society without EJ is not possible, and attempts to attain sustainability without attention to human dignity and justice can only lead to more unequal allocations of environmental bads to parts of the world we may conveniently ignore. At least, that is what has happened up to now. We are – as predicted half a century ago by the Club of Rome – reaching limits to economic growth and natural resource exploitation (Meadows,

Meadows, Randers, & Behrens, 1972). It has been argued that on our current trajectory of population growth and consumer demand, we will soon require the equivalent of three Earths to satisfy our natural resource requirements. That, of course, is simply not going to happen – we inhabit one world, not three. Retrospective work on the predictions of the Club of Rome regarding natural resource scarcity, demographic trends, and per capita industrial output suggests this group of researchers was on target in anticipating the inflection point where ecological resources cannot keep up with incessant economic growth (Turner & Alexander, 2014). That point is now.

Thus, in addition to struggling for the interests of their own communities EJ activists are also reminding the rest of us that the crisis of resource scarcity and toxic exposure in a contaminated world is already here. Moreover, though the environmental impacts of our consumer behavior may seem far away, we are in truth surrounded and infused by them every day. The endocrine disruptors in our "fast fashion" undergarments, the volatile organic compounds in our living rooms, and the carcinogenic amalgam that constitute our transportation vehicles assure that all of us have toxic substances coursing through our veins. Sociologically-speaking, however, we should not be surprised that the detrimental thrust of our resource-intensive lifestyles is being divvied out in an unequal fashion; those with the least economic and political resources in our global society are being hit the hardest first.

Water scarcity and human migration

In the era of climate change, fresh water looks increasingly like the key natural resource capable of effecting changes in populations and geographic mobility everywhere. Climate-induced desertification of arable lands around the world has led to mass migrations of people – principally from rural areas to urban areas in the same country, a demographic phenomenon known as *internal migration* (Rigaud et al., 2018). Recent years have seen media coverage of what are sometimes referred to as "climate refugees," the implication being that global warming has led to a mass exodus of people from the global south to the global north.

As is any human social process, migration is complex, and environmental change is but one element of many – including economic disparities, war, government policies, and household "diversification strategies" – that go into the decision to pack up and live somewhere else. However, there is considerable evidence that more than half of the world population faces fresh water scarcity for at least one month every year. Parts of Mexico, North Africa, South Africa, the Middle East, South Asia, and Australia face water scarcity all year long (Mekonnen & Hoekstra, 2016). Many people whose families and communities have historically depended on agriculture for a livelihood find themselves faced with the prospect of leaving their homelands. When they do, however, their most likely destination is the larger towns and cities in their own country.

The World Bank estimates that by the year 2050 as many as 143 million people will have migrated internally (Rigaud et al., 2018). Global warming, a meteorological shift set in motion over a 200 years ago with the onset of the Industrial Revolution in Europe and North America, is currently effecting its greatest damage among populations with relatively small ecological footprints.

No sustainability without greater social equity

Given the scale of the ecological crisis, the inconvenient truth for green-minded denizens of the global north is that a focus on individual consumer actions is bound to accomplish very little. For example, two dominant approaches for addressing climate change are a) finding technological solutions that can reduce our dependence on fossil fuels in households, industry, and transportation, or b) implementing policies that incentivize behavioral changes in consumption. A justice perspective, however, forces us to consider the disparate impacts technological investments and fiscal incentives such as a carbon tax have on people of varying SES profiles.

French President Emmanuel Macron faced exactly this problem in late 2018 when he decided to implement a green tax on gasoline which disproportionately impacted working-class commuters. "Yellow vest" protesters (named so because they demonstrated wearing the safety vests French drivers are required by law to have in their cars) appeared less opposed to the levying of a carbon tax than to the way it was dealt out – airlines, for example, and cruise ships, though they pollute far more than automobiles, were exempt.

A common critique of Macron during this time is that he was tone-deaf to the needs and fears of everyday French people. To underscore this theme of social distance from the common person, Macron was often caricatured in political cartoons wearing a crown and coronation robe – a low blow in a country whose national slogan is "liberty, equality and fraternity." Similar types of pushbacks against the perceived unjust distribution of ecological austerity can be found at the level of geopolitics. There, countries who have recently experienced rapid economic growth find it unfair that they should be restricted from further industrialization. Europe and the United States, after all, had no such restrictions at the time of their greatest economic expansion in the 19th and 20th centuries.

At the most basic level of political buy-in, an electorate will be reliably uninterested in supporting environmental reform if they perceive that they are the only ones making a sacrifice. Nobody wants to feel they've just been duped – another way of saying social justice is essential for working towards a more ecologically engaged society. Even if their intentions are good, elite decision-makers disconnected from the reality and material needs of average citizens will find their efforts to avert ecological disaster severely hindered. In the next chapter, we will begin to consider some of the more hopeful pathways for working towards a more just, less unequal society, and in doing so, highlight the essential contribution sociology can make to avoiding planetary doom.

Glossary

Bogardus scale   A survey-based scale of social distance, prejudice, and preference for different groups of people in society.

De facto **discrimination**   Discriminatory behavior, though not ordained by law, is carried out in everyday practice, nonetheless.

De jure **discrimination**   Officially sanctioned laws of discrimination and separation on the basis of race that existed prior to the suite of Civil Rights Acts passed by Congress in the 1960s.

Globalization   The economic, social, and cultural impacts on societies due to the movement of industrial production around the world, typically as a strategy to lower the cost of labor and evade regulation.

Housing Covenant   An agreement written into the deed for a home specifying what race/ethnicity the homeowners should be.

Hypersegregation   High levels of residential segregation which concentrates crime and poverty and cuts off opportunity for social mobility and economic advancement across generations.

Internal Migration   Significant population shifts and movement within a state/geographical region, often due to desertification or other environmental issues with longstanding social impacts.

Redlining   The practice of banks discriminating by race who should be given federally subsidized loans to buy a home and who should not, and which neighborhoods they should be allowed to live in.

Social desirability bias   The tendency of survey participants to respond in ways they believe will be perceived positively by researchers.

Underclass   The socioeconomically marginalized individuals and groups segregated by class and race in American society who live in impoverished neighborhoods with underfunded schools and social services, few economic opportunities, high rates of crime and violence, and low-quality housing stock; an idea developed by William Julius Wilson.

White flight   A particular form of housing segregation that took place after World War II wherein white families with ready access to mortgage financing moved to the suburbs, leaving behind a concentration of African Americans and other groups cut off from federally subsidized loans in urban areas.

Questions

1   Bogardus scale surveys suggest that, on average, social distance between different race and ethnic groups in the United States has decreased, meaning we are more tolerant now than we were in the mid-20th century. Social desirability bias, however, suggests this may simply be a matter of people being less willing to express their prejudices openly. What do you think? Are we more or less tolerant as a nation than we were right after World War II?

2   Racism refers to the prejudicial attitudes of individuals towards an entire class of people on the basis of race. *Structural racism*, however, refers to something more insidious, permanent, and debilitating for certain groups of society. Given what you read in this chapter, what can you say about how racism has been structured over time in the United States today, particularly around housing and employment?

3   The critical environmental justice perspective is highly skeptical of the motives of government in creating and enforcing environmental regulations. Activists sharing this point of view believe more direct action is needed and that waiting for the government to implement meaningful changes is useless given the immediate impacts felt by communities living proximate to environment hazards. Yet, the regulation of polluting industries requires a regulatory state, i.e., government intervention. Should environmental justice activists work with or against government?

References

Alexander, M. (2012). *The new Jim Crow: Mass incarceration in the age of colorblindness, 2010*. Revised ed.: New Press.

Badger, E. (2017). How Redlining's racist effects lasted for decades. *The New York Times* (August 24). Retrieved from https://www.nytimes.com/2017/08/24/up-shot/how-redlinings-racist-effects-lasted-for-decades.html. Accessed September 16, 2021.

Bertrand, M., & Mullainathan, S. (2004). Are Emily and Greg more employable than Lakisha and Jamal? A field experiment on labor market discrimination. *American Economic Review, 94*(4), 991–1013.

Bogardus, E. (1926). Social distance in the city. *Proceedings and Publications of the American Sociological Society, 20*(1926), 40–46.

Bogardus, E. (1967). *A forty year racial distance study*: University of Southern California.

Bullard, R. D. (2018). *Dumping in Dixie: Race, class, and environmental quality*: Routledge.

Carter, E. D., Silva, B., & Guzmán, G. (2013). Migration, acculturation, and environmental values: The case of Mexican immigrants in central Iowa. *Annals of the Association of American Geographers, 103*(1), 129–147.

Desmond, M. (2016). *Evicted: Poverty and profit in the American city*: Broadway Books.

Ehrenreich, B. (2010). *Nickel and dimed: On (not) getting by in America*: Metropolitan Books.

Elliott, J. R., & Howell, J. (2017). Beyond disasters: A longitudinal analysis of natural hazards' unequal impacts on residential instability. *Social Forces, 95*(3), 1181–1207.

Elliott, J. R., & Pais, J. (2006). Race, class, and Hurricane Katrina: Social differences in human responses to disaster. *Social Science Research, 35*(2), 295–321.

Florido, A. (2017). Houston's undocumented residents left destitute and fearful in Harvey's wake. *National Public Radio*, National section. 7 September 2017.

Krumpal, I. (2013). Determinants of social desirability bias in sensitive surveys: A literature review. *Quality & Quantity, 47*(4), 2025–2047.

Macias, T. (2016). Environmental risk perception among race and ethnic groups in the United States. *Ethnicities, 16*(1), 111–129.

Massey, D. S., & Denton, N. A. (1993). *American apartheid: Segregation and the making of the underclass*: Harvard University Press.

Meadows, D. H., Meadows, D. L., Randers, J., & Behrens, W. W. (1972). The limits to growth. *New York, 102*, 27.

Mekonnen, M. M., & Hoekstra, A. Y. (2016). Four billion people facing severe water scarcity. *Science Advances, 2*(2), e1500323.

Pager, D. (2003). The mark of a criminal record. *American Journal of Sociology, 108*(5), 937–975.

Pearce, D. M. (1979). Gatekeepers and homeseekers: Institutional patterns in racial steering. *Social Problems, 26*(3), 325–342.

Pellow, D. N. (2017). *What is critical environmental justice?*: John Wiley & Sons.

Pellow, D. N., & Brulle, R. J. (2005). *Power, justice, and the environment*: MIT.

Rigaud, K. K., de Sherbinin, A., Jones, B., Bergmann, J., Clement, V., Ober, K., Schewe, J., Adamo, S., McCusker, B., Heuser, S., Midgley A. (2018). *Groundswell: Preparing for Internal Climate Migration*: World Bank.

Rothstein, R. (2017). *The color of law: A forgotten history of how our government segregated America*: Liveright Publishing.

Sbicca, J., & Myers, J. S. (2017). Food justice racial projects: Fighting racial neoliberalism from the Bay to the Big Apple. *Environmental Sociology, 3*(1), 30–41.

Tessum, C. W., Apte, J. S., Goodkind, A. L., Muller, N. Z., Mullins, K. A., Paolella, D. A., … Marshall, J. D. (2019). Inequity in consumption of goods and services adds to racial–ethnic disparities in air pollution exposure. *Proceedings of the National Academy of Sciences, 116*(13), 6001–6006.

Turner, G., & Alexander, C. (2014). Limits to growth was right. New research shows we're nearing collapse. *The Guardian,* September 2. Retrieved from https://www.theguardian.com/commentisfree/2014/sep/02/limits-to-growth-was-right-new-research-shows-were-nearing-collapse. Accessed September 16, 2021.

Turner, M. A. (2013). *Housing discrimination against racial and ethnic minorities 2012: Executive summary*: Urban Institute.

Wilson, W. J. (2012). *The truly disadvantaged: The inner city, the underclass, and public policy*: University of Chicago Press.

8   Searching for trust in a risk-riddled society

Generalized trust is key to our understanding of why people become engaged in civic life, willing to sacrifice time and resources for the common good. It reflects a positive belief in the good will of others beyond the circle of friends, family, and acquaintances we know on a face-to-face basis (Newton, 1997). Trust grows out of our social integration with others. When we find ourselves in a context where we know many people who also know each other in different ways – as coworkers, neighbors, soccer parents, etc. – we have a solid basis for building trust. If this is a context we have grown up in and we have a relatively diverse set of network contacts, our place-specific trust can foster an even broader sense of generalized trust towards those we have not even met.

The dark side of too much trust

That said, we probably do not want to get too comfortable with the notion that trust is the solution to all our problems. The literature on social capital, for example, reminds us that collective action based on trusting social bonds does not always result in positive outcomes. Organized crime, pyramid schemes, and the in-group exploitation of immigrant groups where ethnic bonds can sometimes substitute for fair pay and decent working conditions all point to the potential "dark side" of trust where perhaps a degree of mistrust is in order (Portes, 1998; Portes & Sensenbrenner, 1993; Van Deth & Zmerli, 2010). The omnipresence of digital media may make us also want to be extra cautious about the intentions of others. Be it the personal information that can be extracted from our online accounts, or the possibility of malware tracking our every click and keystroke, many of us have now been socialized into a state of perpetual electronic vigilance. Generalized trust in this particular case looks increasingly like a sucker's game.

If our direct personal ties to each other are not always used for mutual benefit, it stands to reason that generalized trust in others may not necessarily result in collectively positive outcomes. Where, then, do we draw the line? Do we simply throw our hands up and carry on as self-interested individuals whose best source of security in an untrusting world is data encryption and

DOI: 10.4324/9781003110668-8

a very good password? Or, can we as social scientists try to get an empirical handle on when trust is good and bad, and under what circumstances are people willing to trust each other enough to work for a common good? As we shall see in this chapter, this is important because in the face of mounting socioecological crises, evidence suggests that trust in others is one of the best predictors that people are willing to change behavior and support policies to mitigate global calamity.

The case for generalized trust

But first, what exactly do we mean by *generalized trust*? Beyond the personal trust we have in people we interact with on a regular basis, generalized trust may be thought of as a widespread trust in the integrity of others we do not yet know (Brunie, 2009; Uslaner, 2008). The key point here is that generalized trust among individuals is associated with positive altruistic outcomes, including volunteering, giving to charity, moderation, and self-sacrifice (Brehm & Rahn, 1997; Uslaner, 2008). In his classic study on Italian democracy, Robert Putnam found that in the northern regions of Italy where people had relatively high rates of trust in each other, more extensive network ties, and a long history of civic participation, democracy thrived. In contrast, in southern Italy where trust resided mainly in family networks and civic participation was low, corruption was rampant, people had little confidence in government and democracy withered (Putnam, Leonardi, & Nanetti, 1994).

In the environmental literature it has been argued that trust makes it more likely that people will actively engage in collective sustainable behavior as it builds confidence others will behave in a similar fashion (Pretty & Ward, 2001). People, as it turns out, are strongly influenced by the perceived attitudes and behavior of those around them. In part, we all need reassurance that the changes we make are worth it. Why should I sort out all my trash for recycling if no one else is doing it?, one might reasonably ask. As it happens, a series of studies in "norm activation theory" in social psychology demonstrate that simply observing other people's willingness to recycle is enough for most people to engage in this behavior, provided they think it is a good idea (Derksen & Gartrell, 1993).

Even without directly observing certain kinds of behavior, people can be swayed. In a finding that seems almost silly, the positive reinforcement of a smiley face on electric bills for successfully reduced energy consumption has been shown to have a significant impact on conservation behavior (Schultz, Nolan, Cialdini, Goldstein, & Griskevicius, 2007).

Trust thus serves as an important link between attitudes and behavior – not a minor point since the difference between what people say and what they actually do has long been the bane of attitudinal research. This is especially true for environmental attitudes, which is sadly a little too much like the old joke about the weather: Everybody complains; nobody does anything about it (attributed to Charles Dudley Warner, though Mark Twain liked to quote him on this). It is not that people are lying when they take environmental surveys, but it is clear there are enormous obstacles in the way of carrying out environmentally conscious beliefs in the real world.

Confronted with the massive die-off of the world's species, most people would prefer something be done to stop it. Unfortunately, the sense that one person's actions would probably make little difference is probably true. That said, a large group of people with a common purpose have been shown to change the course of history, and it is specifically in that context that trust in others becomes crucial. In addition to providing the basis for social movement organization, higher levels of trust among individuals have been linked to an array of sustainable practices, including water conservation (Van Vugt & Samuelson, 1999), recycling (Mannemar Sønderskov, 2011), the use of public transportation (Van Lange, Vugt, Meertens, & Ruiter, 1998), and the consumption of "green" products (Gupta & Ogden, 2009).

The source of generalized trust is itself the basis of much debate. A directly structural argument would have it that generalized trust is simply a result of socioeconomic advantage, indicative of the self-confidence and faith in others associated with positions of privilege (De Hart & Dekker, 2003). Indeed, research on social capital shows higher levels of civic engagement and trust among people of greater social status, as regards, income, education, and occupation (Putnam, 2000).

A somewhat more nuanced perspective would acknowledge class differences while underscoring the significance of participation in community activities for nurturing generalized trust among individuals across the socioeconomic spectrum (Brehm & Rahn, 1997; Stolle & Rochon, 1998). And, while disadvantaged populations may exhibit lower levels of generalized trust than those in positions of relative power, face-to-face interactions in diverse neighborhood settings appear to be an important predictor of this aspect of social capital among the poor and race and ethnic minorities in the United States and Canada (Marschall & Stolle, 2004; Stolle, Soroka, & Johnston, 2008).

These findings are consistent with the social network research on multiplexity referenced in Chapter 3 – regular interaction with the same people in different contexts is an excellent source of interpersonal trust. As pointed out in Chapter 5, regular interaction among neighbors may also help lessen the distorting influence of advertising and social inequality on consumption (Schor, 1999).

The case against trust – environmental injustice

Before we take a ride down the slippery slope of general trustiness, we should consider a central challenge to this approach within environmental sociology. Namely, the environmental justice literature which, not incidentally, places special emphasis on the unequal distribution of environmental risks and political influence in American life. As has already been pointed out, generalized trust tends to concentrate disproportionately among those in relative positions of power. Well-paid, highly educated whites tend to have, all things considered, higher levels of generalized trust than those not matching that profile (Putnam, 2007).

Sociological research in environmental justice, however, suggests we look beyond the mere question of social status. For example, an important correlate of lower levels of generalized trust among people of color is exposure to environmental hazards, such as air pollution, agricultural pesticides, and lead poisoning, to name a few (Adeola, 2004; Jones & Rainey, 2006). Thus, it is not simply that higher status affords some people a greater sense of trust because of their superior social contacts; at the lower end of the scale, exposure to actual risks, including environmental hazards, has an impact on people's ability to trust. Put another way, if it is clear that those in positions of power do not care about the quality of the air, water, housing, and environment where I live, why should I trust them?

In the 1980s, starting from a point of deep skepticism towards those in power, communities of color in the United States began generating awareness of environmental injustices where they lived through place-based activism and strategies closely aligned with the rights and justice model of the Civil Rights Movement (Bullard, 1990; Bullard, Mohai, Saha, & Wright, 2007; Pellow & Brulle, 2005). Inspired by these early protests, sociological research began to map out race and ethnic differences in environmental concern. A theme that emerged is that, though whites had on average expressed greater concern for global environmental themes such as climate change and ozone depletion, Blacks and Latinos often express greater concern than whites for risks that might directly affect their health or that of the local community (Arp III & Kenny, 1996; Jones & Rainey, 2006).

Apart from attitude surveys, sociological research has made clear the empirical link between the race and class distribution of people in American cities and their actual exposure to health risks generated by industry and the disposal of toxic waste (Bullard et al., 2007; Jones, 1998). It is apparent that racial differences in environmental concern are tied to a group's sense of environmental threat and, when environmental risks are included as a central measure of environmental concern, whites often show less concern than non-whites (Mohai & Bryant, 1998). In addition, persistent race-based differences in unemployment, the greater likelihood of violence in minority communities, and disproportionate incarceration rates for Black and Latino men suggest race and ethnic differences in generalized trust are strongly tied to the unequal distribution of risks and opportunity in modern life (Massey, White, & Phua, 1996; Ross, Mirowsky, & Pribesh, 2001).

Is trusting others really worth it?

The implications of generalized trust for environmental outcomes are thus two-pronged. On the one hand, generalized trust should predict a willingness to sacrifice for the environment in terms of lifestyle adjustments – using alternative forms of transportation, recycling, reducing household energy and water consumption – and even a willingness to pay higher prices and taxes in the name of environmental protection. These kinds of changes are consistent with the underlying theme that trust in others is a great basis for people to work for the common good (Bridger & Luloff, 2001; Stolle & Rochon, 1998).

On the other hand, generalized trust may also be associated with a kind of ecological naïveté or faith that business-as-usual practices in industry, government, and consumer markets effectively address environmental threats. It would be easy to imagine that among a class of people who are not regularly exposed to environmental hazards, business-as-usual would be the way to go. We can expect, and the literature predicts that, when greater trust leads to diminished concern for environmental threats, the willingness to make sacrifices for the environment will diminish as well (Dietz, Stern, & Guagnano, 1998).

My own research using national data from the General Social Survey or GSS (Macias, 2015) shows support for both hypotheses. In the end, however, the direct positive effect of generalized trust on a willingness to make sacrifices for the environment outweighs its indirect negative effect on this same variable via the perception of environmental risks by a ratio of 4 to 1. That is, as relates to addressing environmental issues, trust in others is worth it. A healthy skepticism of others' intentions as evidenced through environmental justice activism is also good, but in the final tally to attain collective socioecological goals, trust in others is better than not trusting at all.

This finding comes, however, with a significant caveat: To the extent generalized trust is predicated on having socioeconomic advantage, we should be skeptical of efforts to promote greater trust which do not address social inequality. In fact, three of the most significant background predictors of generalized trust in my own research are education, income, and race (whites are on average more trusting than non-whites). Recent concerns raised about the significance of racial diversity vis-à-vis trust in American communities (Putnam, 2007) should only underscore the need to address social inequality if we wish to also formulate and attain common goals.

In addition, the single most influential variable in my modeling of GSS data is the sense of being directly affected by environmental problems. As environmental justice activism suggests, the perception of environmental threat is a powerful undercurrent capable of drawing people towards a willingness to make sacrifices for the sake of socioecological change.

Sense of threat and political activism in Flint, Michigan

A recent example of this can be found in the case of Flint, Michigan where changing the municipal source of water from Lake Herring to the Flint River resulted in lead poisoning among much of the population. Flint residents were suddenly confronted with the direct impact of water contamination brought on by the poor planning of regional leaders. Though at first "water warrior" activists relied primarily on scientists to back up their claims of environmental injustice, it soon became apparent that the more essential problem they faced was one of representative democracy. That is, "emergency managers" who had been put in place to handle the crisis were not very interested in addressing community grievances (Pauli, 2019).

The Flint case demonstrates that technocratic expertise can at times be used to silence the voices of those most directly threatened by environmental hazard. Environmental justice activists thus have reason to be leery of outside expertise. Though science is an essential tool in making a persuasive empirically-based argument, it is the political confrontation with those in power that usually proves most effective in combating injustice.

Our distorted sense of risk

Most people, however, do not share a direct sense of threat from looming environmental risks even though most reliable scientific research suggests we will all be impacted by the rapid changes taking place in the Earth's ecosystems. Where, then, will our sense of risk come from? Moreover, even if we can answer that question, is there any guarantee we will be motivated to act? Despite its link to statistical analysis and its air of mathematical authority, our perception of risk often rests on a pretty wobbly foundation.

This may come as a surprise given the insurance industry – 13 percent of US GDP – is utterly dependent on the calculation of risk. The gaming industry – a mere 1 percent of US GDP – is also highly dependent on the precise calculation of probability and the rather imprecise, often bizarrely optimistic calculations of its clientele. Unfortunately, in the realm of public opinion, most of us have more in common with the naïve gambler who has had perhaps one too many complementary drinks than the average casino owner.

Not surprisingly, the social sciences have taken a variety of approaches to the question of risk that often reflect the particular proclivities of each discipline. Within economics, *uncertainty* refers to random events that occur outside of our control. We are thus surrounded by uncertainty which can make most of us a little anxious. *Risk*, by contrast represents our corralling to the best of our ability the knowledge we have to in some sense "control" uncertainty by generating an estimate of probability. So, if you have a little experience under your belt and not too many unsolicited raspberry mojitos, you know the best gambling odds in the casino on the edge of town are at the Black Jack tables. The odds are still stacked against you, of course, but if you can count to 21 you might be able to gain back some of what you lost at the slot machines.

Hopefully, a more common sense of risk for the reader comes when making a habitual check on a smartphone weather app. There we mainly understand that a "30 percent chance of rain" does not mean it will be raining 30 percent of the day. That said, even with a decent understanding of how probability is calculated, I still occasionally need to remind myself: This same forecast implies the inverse – a 70 percent chance of no rain. As any casino mogul or meteorologist will tell you, risk is calculable. Unfortunately, most of us have neither the necessary data nor computing savvy to outdo the pros. Tragically, even with good information at our disposal, we often deceive ourselves into making bad decisions. What's that about? Fortunately, sociology has some answers.

Ulrich Beck's *Risk Society*

With the publication of his book, *Risk Society* (1992), sociologist Ulrich Beck became one of the most influential thinkers on the topic of modern day risk. His work was in part inspired by the increasing prevalence of environmental disasters such as the 1986 Chernobyl nuclear accident in the Ukraine. At that point, it was becoming evident that modern life, along with providing amazing comforts and convenience – the "goods" – also came with a lot of "bads." In many ways, what Beck is accounting for here is a major shift in the scale of risk in complex technological societies.

At the beginning of the 20th century, Americans were still a predominantly rural society, subject to the elements, reliant on the Farmer's Almanac to reduce risks and best predict the coming agriculture year. Contagious diseases such as measles and tuberculosis were perpetually lethal threats, and the rate of mothers dying in childbirth was about 40 times greater than it is today. At the individual level, there is no question; we live longer on average, and have safer lives than our predecessors did 100 years ago. Though we can be nostalgic about the past, most people do not really want to go back to the time before modern medicine, indoor plumbing, and near universal electrification, to say nothing of contemporary forms of transportation and telecommunication.

Despite, or perhaps because of all these comforts and conveniences, this modern life we inhabit is, according to Beck, "reflexive." By that, he does not mean we reflect deeply about the consequences of our actions – far from it. Instead, Beck observes that as society races towards greater complexity and an ever-greater dependence on technological innovation, we often discover after the fact that much of this innovative disruption comes at a cost.

Comfortable, heated homes, airplanes, and the internet all require massive amounts of energy and, despite remarkable advances in technological efficiencies, we continue to consume more and more energy every year. One of the main byproducts of fossil fuel consumption – still by far the greatest source of global energy production – is greenhouse gas emissions, which have also increased in magnitude every year since the beginning of the industrial revolution, roughly two centuries ago.

Perhaps you feel, as many have argued, that nuclear power provides the "clean" alternative to dirty fossil fuels which, along with heating up the planet, have directly contaminated pristine marine ecosystems – see the discussion of the Exxon Valdez and Deepwater Horizon disasters in Chapter 2 – and have been the driving force behind ocean acidification. Nuclear power, however, generates its own radioactive waste stream which, as also mentioned in Chapter 2, presents daunting challenges that have yet to be resolved. Moreover, the existential risks presented by nuclear melt-downs such as those at Chernobyl in 1986 and Fukushima in 2011, and the ongoing though sometimes forgotten threat of nuclear holocaust posed by the worldwide stock of armed nuclear warheads get to the essence of what Beck means by living in a risk society. Though, relative to the past, we may feel safer as individuals, our collective future as lifeforms on the planet seems more at risk than ever before.

The perils of "reflexive management"

Even so, the fact that contemporary risks appear much larger in scale than in the past makes it difficult to put our finger on how they directly affect us. Climate change, for example, is an enormous planetary hazard set to affect us all, but it is still, relative to a human lifespan, a gradual change, making it easy for cynical politicians to point to the latest snowstorm as evidence that there is nothing to worry about. What hope, then, do well-intentioned poli-cymakers have for persuading a skeptical public?

In that regard, Beck's key recommendation – and the title of another book of his – is "reflexive management." That is, government no longer has con-trol over information available to the public. The public may have easy access to an arsenal of information with which to challenge unwelcome policy deci-sions. If government officials are to have any hope of implementing successful policy, they must (a) listen and anticipate what the affected public has to say, and (b) think long and hard about how to best craft a palatable message.

This last point implies that government should be at least as concerned about the use of emotional appeals as rational points of fact in the formulation of effective messaging. The danger here in advocating for risk management is that, once emotional appeals begin to outweigh points of fact, pretty much anything goes. Appealing to people's pre-existing bias and prejudice may be an expedient way to get votes, but has horrendous implications for represent-ative democracy.

Terrorism, for example, is a heinous ideologically driven act of violence inflicted on innocent lives that must be condemned wherever it occurs. As actual risks go, however, it is also exceedingly rare. In an average year in the United States, you are about 25 times more likely to die in an aviation acci-dent (another relatively rare event), and almost 2,500 times more likely to die in a car accident than die in a terrorist attack (Nowrasteh, 2016). Nonetheless, in 2015, a year before the US presidential election, 47 percent of Americans worried their family members would be victims of terrorism (Swift, 2015). Risk among the general public, as it turns out, is mainly a question of percep-tion which can be easily manipulated.

For example, when politicians deliberately link terrorism to an entire class of immigrants they are engaging in a kind of risk management utterly devoid of meaningful evidence. In the United States, immigrants as a group commit less crime than native-born Americans, and there is even some evidence that a greater presence of immigrants in neighborhoods is associated with greater safety (Ewing, Martinez, & Rumbaut, 2015; Light & Miller, 2018). Nonetheless, there is a broad perception among the Americans that immigration is directly linked to terrorism even though it is not. The vast majority of terrorist acts in the United States have been committed by native-born citizens.

Even in countries like France and Belgium which have sadly seen an uptick in terrorist attacks over the past decade, terrorism is typically an act committed by European citizens. Broader issues such as social exclusion and a lack of economic opportunity for the younger generation appear as key conduits to violence and radical ideologies that reject the dominant society (Khosrokhavar, 2016).

Changing perceptions of the risk of immigration

Using immigration as a theme, let's examine briefly how a particular risk has been distorted in recent decades and what might be done to lessen the sense of threat among the general public. First, it might be helpful to look back at a time when the lack of immigration was seen as one of the greatest threats to American economic growth and technological knowhow. On October 4, 1957, the Soviet Union famously launched the first human-made satellite into orbit around the Earth. All in all, an amazing accomplishment for the human species, though that is not how it was viewed by the US government.

The launch of Sputnik was near the height of the Cold War between the United States and the USSR. The nuclear arms race between these two nations was well under way and there was a pervasive paranoia – a heightened sense of risk also manipulated by those in power – around the potential infiltration of communist values into American culture. Along with the question of who was a loyal American and who was not, it was becoming evident that the growing mass of middle-class students entering college was not sufficiently interested in math and science in numbers that would challenge the dominance of the USSR in those areas. Something had to be done.

The Hart-Celler Act

One remarkable strategy required changes in federal immigration law. Through the first half of the 20th century, US immigration preferences were largely based on race, restricting entry from most of the Asian continent and giving strong preferences to people from northern Europe. All that changed in 1965 with the passage of the Hart-Celler Act which, though establishing country and regional level quotas, opened up migration to the United States from pretty much everywhere around the world. There were still preferences in the new legislation, but they were now based principally on pragmatic and humanitarian reasoning. On the pragmatic side, the Hart-Celler Act gave strong preferences to people with needed skills and education in math and science, regardless where they came from. Not surprisingly, Asia, which had long been a restricted region in federal immigration law, saw a large number of its technically skilled, highly educated workers enter the United States after 1965.

On the humanitarian side, the new immigration laws also gave a strong preference for family reunification. This meant that not only could immigrant-origin American families sponsor their spouses and children to enter the country with legal resident status, they could also sponsor parents and even siblings to do so. In brief, the Hart-Celler Act recognized that immigrants are not simply abstract units of labor employed in the economy; they are also humans in need of affective, social, and emotional support, and maintaining intact families is an efficacious way of providing these non-economic elements. The Hart-Celler Act also gave a lower ranking preference for refugees, a category that would be more clearly addressed in the Refugee Act of 1980. Again, the view of risk at this point vis-à-vis immigration was that the greatest risk would be neglecting the pragmatic and humane management of this social phenomenon.

The United States' approach to immigration management dramatically changed in the mid-1990s. At that point, the United States was coming out of a years-long economic recession. The general trend in federal legislation at the time was towards the privatization of government services and the de-regulation of industry; maneuvers that were seen as cost-cutting in the public sector and profit-generating in the private sector. Neither strategy, however, came with an inherent base of popular political support.

In the early 1980s, French philosopher Michel Foucault wrote and lectured about what he termed *governmentality* (Foucault, Davidson, & Burchell, 2008). By this he meant the tendency of those in power to attempt to create the kinds of subjects or citizens they need during particular moments in history. In the mid-1990s, one could argue that the US government was attempting to change the meaning of what it meant to be an immigrant. Whereas during a time of great economic expansion during the 1960s immigrants were seen as an asset necessary to further our collective progress as a nation, by the mid-1990s, they were a scapegoat; an explanation for unemployment and the lack of socioeconomic mobility among working-class Americans.

"Illegal" immigrants and governmentality

On September 30, 1996, President Bill Clinton signed into law the *Illegal Immigration Reform and Immigrant Responsibility Act* (IIRIRA). Passed at a time of historically high rates of undocumented immigration from Mexico, IIRIRA – in direct contrast to the Hart-Celler Act three decades prior – was highly vindictive. It provided funding for the construction of border walls, the hiring of an additional 1,000 border patrol agents, the beginning of a trend towards the mass deportation of undocumented immigrants, as well as the incarceration of "illegal" immigrants in private for-profit prisons.

Beyond the convenience of blaming a class of politically defenseless people for the nation's economic woes, the increasing use of the term "illegal" in political discourse – it is in the title of the IIRIRA – worked to create a category of workers essential and highly profitable for certain sectors of the economy. It is estimated, for example, that 26 percent of agricultural workers are undocumented (Passel & Cohn, 2016). Hospitality, restaurants, and domestic cleaning services are also known to rely heavily on this group which, living under the constant threat of incarceration and deportation, is unlikely to complain about low wages, long work hours, or arduous working conditions. In addition, the three quarters of all undocumented prisoners housed in private prisons has turned immigrant detention into a multi-billion dollar industry (Gilman & Romero, 2018).

From the point of view of Foucault's governmentality, one might argue that undocumented immigrants serve the function of both convenient political distraction and, in some industries, a reliable source of profit. Combine this with the irrational fears of terrorism tied to the public's perception of immigrants more broadly, and it is not difficult to see how immigration has become one of the most contested and politically divisive topics in the country, if not the world, today.

Fear of immigrants is one of the greatest threats to our collective sense of trust and common purpose. One might reasonably ask, given the divisiveness and demographic changes already at play, can we ever learn to trust each other as a nation again? Fortunately, as we will see in the following chapters, not all hope is lost. In fact, decades of research suggests the key to letting down our guard towards people different from ourselves is simply taking the time to know them.

Glossary

Generalized trust A positive belief in the good will of others beyond the circle of people we know on a face-to-face basis; associated with positive altruistic outcomes.

Governmentality The tendency of those in power to attempt to create the kinds of subjects or citizens they need during particular moments in history.

Reflexive management A strategy taken by government officials to anticipate public opinion so as to best craft a palatable message that persuades their constituents of needed policy changes for the common good.

Risk An estimate of probability that utilizes to the best of our ability the knowledge we have to, in some sense, "control" uncertainty.

Risk society Ulrich Beck's idea that though we may lead safer more comfortable lives than in the past, our collective future seems more at risk than ever before.

Uncertainty Random events that occur outside our control.

Questions

1   "Generally speaking, would you say that most people can be trusted, or that you can't be too careful in dealing with people?" How would you respond to this question regularly asked in the GSS? How do you think most people in American society would respond to it? What are the political consequences of promoting trust versus distrust? Do you believe at this point in time we as a society are leaning more towards general trust or general distrust?

2   Ulrich Beck argues that the greater comfort and convenience we experience in modern life correlates with exposure to more disruptive manmade risks that spiral seemingly out of our control. Fair enough, but is the inverse true, too? Could we reduce broader societal risks by letting go of a few of our unnecessary conveniences and comforts? Can you think of a few examples of this and their implications for evading global calamity?

3   In this chapter, we talked about Foucault's notion of governmentality and how the role of immigrants in American society has been redefined to fit the needs of society at different points in time. Can you think of another social category that has also been redefined in this way? Who benefited from this redefinition? How were the people who occupied this category affected?

References

Adeola, F. O. (2004). Environmentalism and risk perception: Empirical analysis of black and white differentials and convergence. *Society & Natural Resources, 17*(10), 911–939.

Arp III, W., & Kenny, C. (1996). Black environmentalism in the local community context. *Environment and Behavior, 28*(3), 267–282.

Beck, U., Lash, S., & Wynne, B. (1992). *Risk society: Towards a new modernity* (Vol. 17): Sage.

Brehm, J., & Rahn, W. (1997). Individual-level evidence for the causes and consequences of social capital. *American Journal of Political Science, 41*, 999–1023.

Bridger, J. C., & Luloff, A. E. (2001). Building the sustainable community: Is social capital the answer? *Sociological Inquiry, 71*(4), 458–472.

Brunie, A. (2009). Meaningful distinctions within a concept: Relational, collective, and generalized social capital. *Social Science Research, 38*(2), 251–265.

Bullard, R. D. (1990). *Dumping in dixie*: Westview.

Bullard, R. D., Mohai, P., Saha, R., & Wright, B. (2007). *Toxic wastes and race at twenty 1987–2007: Grassroots struggles to dismantle environmental racism in the United States*: United Church of Christ Justice and Witness Ministries.

De Hart, J., & Dekker, P. (2003). A tale of two cities: Local patterns of social capital. In M. Hooghe & D. Stolle (Eds.), *Generating social capital: Civil society and institutions in comparative perspective* (pp. 153–169): Palgrave.

Derksen, L., & Gartrell, J. (1993). The social context of recycling. *American Sociological Review, 58*, 434–442.

Dietz, T., Stern, P. C., & Guagnano, G. A. (1998). Social structural and social psychological bases of environmental concern. *Environment and Behavior, 30*(4), 450–471.

Ewing, W. A., Martinez, D., & Rumbaut, R. G. (2015). *The criminalization of immigration in the United States.* American Immigration Council Special Report.

Foucault, M., Davidson, A. I., & Burchell, G. (2008). *The birth of biopolitics: Lectures at the Collège de France, 1978–1979*: Springer.

Gilman, D., & Romero, L. A. (2018). Immigration Detention, Inc. *Journal on Migration and Human Security, 6*(2), 145–160.

Gupta, S., & Ogden, D. T. (2009). To buy or not to buy? A social dilemma perspective on green buying. *Journal of Consumer Marketing, 26*(6), 376–391.

Jones, R. E. (1998). Black concern for the environment: Myth versus reality. *Society & Natural Resources, 11*(3), 209–228.

Jones, R. E., & Rainey, S. A. (2006). Examining linkages between race, environmental concern, health, and justice in a highly polluted community of color. *Journal of Black Studies, 36*(4), 473–496.

Khosrokhavar, F. (2016). *Prisons de France: violence, radicalisation, déshumanisation: surveillants et détenus parlent*: Robert Laffont.

Light, M. T., & Miller, T. (2018). Does undocumented immigration increase violent crime? *Criminology, 56*(2), 370–401.

Macias, T. (2015). Risks, trust, and sacrifice: Social structural motivators for environmental change★. *Social Science Quarterly, 96*(5), 1264–1276. doi:10.1111/ssqu.12201

Mannemar Sønderskov, K. (2011). Explaining large-N cooperation: Generalized social trust and the social exchange heuristic. *Rationality and Society, 23*(1), 51–74.

Marschall, M. J., & Stolle, D. (2004). Race and the city: Neighborhood context and the development of generalized trust. *Political Behavior, 26*(2), 125–153.

Massey, D. S., White, M. J., & Phua, V.-C. (1996). The dimensions of segregation revisited. *Sociological Methods & Research, 25*(2), 172–206.

Mohai, P., & Bryant, B. (1998). Is there a "race" effect on concern for environmental quality? *Public Opinion Quarterly, 62*(4), 475–505.

Newton, K. (1997). Social capital and democracy. *American Behavioral Scientist, 40*(5), 575–586.

Nowrasteh, A. (2016). Terrorism and immigration: A risk analysis. *Cato Institute Policy Analysis*, 798, September 13. Retrieved from https://www.cato.org/policy-analysis/terrorism-immigration-risk-analysis. Accessed September 16, 2021.

Passel, J. S., & Cohn, D. (2016). Size of US unauthorized immigrant workforce stable after the Great Recession. *Pew Research Center, 41*, 41.

Pauli, B. J. (2019). *Flint fights back: Environmental justice and democracy in the Flint water crisis*: MIT Press.

Pellow, D. N., & Brulle, R. J. (2005). *Power, justice, and the environment*: MIT Press.

Portes, A. (1998). Social capital: Its origins and applications in modern sociology. *Annual Review of Sociology, 24*(1), 1–24.

Portes, A., & Sensenbrenner, J. (1993). Embeddedness and immigration: Notes on the social determinants of economic action. *American Journal of Sociology, 98*(6), 1320–1350.

Pretty, J., & Ward, H. (2001). Social capital and the environment. *World Development, 29*(2), 209–227.

Putnam, R. D. (2000). *Bowling alone: The collapse and revival of American community*: Simon and Schuster.

Putnam, R. D. (2007). E pluribus unum: Diversity and community in the twenty-first century the 2006 Johan Skytte Prize Lecture. *Scandinavian Political Studies, 30*(2), 137–174.

Putnam, R. D., Leonardi, R., & Nanetti, R. Y. (1994). *Making democracy work: Civic traditions in modern Italy*: Princeton University Press.

Ross, C. E., Mirowsky, J., & Pribesh, S. (2001). Powerlessness and the amplification of threat: Neighborhood disadvantage, disorder, and mistrust. *American Sociological Review, 66*(4), 568.

Schor, J. (1999). *The Overspent American: Why we want what we don't need*: HarperCollins.

Schultz, P. W., Nolan, J. M., Cialdini, R. B., Goldstein, N. J., & Griskevicius, V. (2007). The constructive, destructive, and reconstructive power of social norms. *Psychological Science, 18*(5), 429–434.

Stolle, D., & Rochon, T. R. (1998). Are all associations alike? Member diversity, associational type, and the creation of social capital. *American Behavioral Scientist, 42*(1), 47–65.

Stolle, D., Soroka, S., & Johnston, R. (2008). When does diversity erode trust? Neighborhood diversity, interpersonal trust and the mediating effect of social interactions. *Political Studies, 56*(1), 57–75.

Swift, A. (2015). Americans more worried about terrorism than mass shootings. *Gallup*, December, 16.

Uslaner, E. M. (2008). Where you stand depends upon where your grandparents sat: The inheritability of generalized trust. *Public Opinion Quarterly, 72*(4), 725–740.

Van Deth, J. W., & Zmerli, S. (2010). *Introduction: civicness, equality, and democracy—a "dark side" of social capital?*: Sage.

Van Lange, P. A., Vugt, M. V., Meertens, R. M., & Ruiter, R. A. (1998). A social dilemma analysis of commuting preferences: The roles of social value orientation and trust 1. *Journal of Applied Social Psychology, 28*(9), 796–820.

Van Vugt, M., & Samuelson, C. D. (1999). The impact of personal metering in the management of a natural resource crisis: A social dilemma analysis. *Personality and Social Psychology Bulletin, 25*(6), 735–750.

9 Unstable structures and the process of socioecological change

In fall 2018, France experienced torrential rains with major flooding along the French Riviera and on the island of Corsica. The city of Marseille, where I lived at the time, also received unusually high amounts of rain, though flooding was not as severe or debilitating as it was in other parts of the country. This sense of having averted disaster came to an end, on November 5 when two residential buildings in Marseille's Noailles district suddenly collapsed, killing eight people. Among the dead were immigrants from Algeria, Tunisia, and Comoros, and an Italian graduate student in economics (Chrisafis, 2019). The immediate explanation for what had transpired was that near record levels of rainwater had compromised the integrity of these two 18th-century structures. True enough, however, it did not take long for residents to piece together patterns of neglect and preferential treatment in this Mediterranean metropolis (France's second largest city) that favored commerce and tourism while relegating low-income and immigrant origin residents to dilapidated housing.

In the aftermath of the Noailles tragedy, city inspectors deemed 111 apartment units uninhabitable, requiring over a thousand people to be evacuated from their homes. This was not simply the result of recent downpours. An earlier 2015 report found that over 40,000 units in Marseille were uninhabitable, affecting the safety of over 100,000 residents (Chrisafis, 2019). Nothing had been done since that time to reduce these risks. A lack of municipal funding seemed an obvious explanation for not reinforcing and buttressing this quantity of at-risk residential buildings. However, money was available for other aspects of development. Over the previous decade, 7 billion euros had been spent to renovate the city's waterfront which now includes a high-end shopping district and a chic Museum of European and Mediterranean Civilizations (MUCEM). The latter's unveiling coincided with Marseille's one-year reign as the European Capital of Culture, bestowed to it by the European Union in 2013 (Wainwright, 2013).

The building collapse in Noailles exemplified for many working-class people in Marseille extreme inequality in the treatment of this city's residents. While enormous amounts of resources were put into promoting tourism, cleaning up the historic Old Port, and making way for luxury hotels,

DOI: 10.4324/9781003110668-9

neighborhoods adjacent to economic development projects were largely neglected. Some housing rights activists suspected low-income residents were being abandoned as part of a broader strategy to free up real estate for more high-stakes development. Distrust of government elites and politicians seemed cemented in February 2019 when an investigative report by a local newspaper revealed that some lawmakers were themselves renting out decrepit downtown living spaces at inflated prices (Chrisafis, 2019).

Wherever you look in the world these days, it seems the poor and the working class are increasingly skeptical about the motives of those with greater economic and political power than themselves. Political movements on the right and the left appear to be in part driven by an underlying sense that the status quo is unjust and can no longer stand. New political parties and extremists political candidates reflect the sense among many in the voting public that there is no longer a middle ground to stand on, an understandable point of view given the vast economic divide between rich and poor.

As the example from Noailles underscores, trust is one of the first casualties of social inequality. When representative democracy appears to work in the favor of only a small minority who can disproportionately affect political outcomes through their wealth and influence, faith in the political process erodes. Uncertain times, especially for those at the bottom of the economic ladder, are distressing. They also, however, provide society with the greatest opportunities for social change. Case in point; in the wake of political uproar and organization that followed the Noailles building collapse, the people of Marseille ousted corrupt government officials and elected its first Green Party (*Les Verts*) mayor – a woman, also a first – in summer 2020 (Leras, 2020).

The direction of change – for the collective good or the collective bad – largely depends on levels of generalized trust in society. Broadly speaking, the more we as a society trust each other, the greater the likelihood we can come together as citizens to solve our common problems. When elected officials see high levels of solidarity among the general public, they too are more likely to pay heed and address common grievances.

Inequality: the social distance machine

Unfortunately, the inverse also appears to be true – political opportunists will do their best to encourage and exploit divisions within class, race, gender, religion, and other social categories if it will work to their advantage. Where there is a lack of trust, fear, and hatred along categories of difference will often step in to fill the vacuum. As discussed in an earlier chapter, social distance goes a long way in explaining levels of trust among different groups in society. In his book *Heterogeneity and Inequality* (1977), Sociologist Peter Blau described the way inequality works as a kind of social distance machine. In a relatively equal society where wealth differences are not so great, where public schools and church services are filled with children and families from a variety of socioeconomic backgrounds, we can expect higher levels of interaction across categories of differences. By contrast, in a society where increasingly people congregate with people similar to themselves, attend class and race segregated schools, and in general stay within their cultural comfort zone, there will be fewer opportunities for cross-category interaction and lower levels of generalized trust.

If that last description sounds vaguely familiar, it is probably because it lines up directly with the definition of homophily we covered in Chapter 6, i.e. the tendency of people to befriend and affiliate with people like themselves, a trend only amplified by the use of online media and software-based social networks. We are arguably more selective of our social ties than we have ever been before. As Peter Blau would predict, trust within groups of similar people should be very high. Unfortunately, this comes at the cost of intergroup trust between groups. When the socioeconomic differences among groups are great, as they tend to be in a highly unequal society, levels of generalized trust wither on the vine.

Given current trends, is it even possible to imagine a more equal, trusting society? Put another way: can the topsoil of social integration, depleted of diverse life strategies and complex interdependent ties outside the dominant pull of profit-seeking and consumer culture, ever be revitalized? Fortunately, recent history provides us with a number of examples where social norms and politics that were once considered immutable were called into question and became mutable. In that regard, consider the experience of African Americans – 246 years of slavery and an additional century of enforced segregation and discrimination maintained Black people as *de jure* second-class citizens.

The political process of social change

Something happened, though, in the post WWII era which led to rapid social and political change. The common experience of oppression among African Americans during slavery and after abolition is an obvious starting point. In 1950, over 50 percent of African Americans lived in poverty. As described in Chapter 7, housing discrimination was overt and systematically kept Blacks away from federally subsidized loans and segregated whites-only neighborhoods. As sociologist James Loewen (2018) has written, entire towns across the United States enforced "sundown" bans which prohibited Blacks from staying in these places or even walking on their streets after nightfall. Political participation though legally possible, was in many parts of the country, a farce for African Americans. Literacy tests, polls taxes, and the threat of violence effectively kept Blacks away from voting sites (Morris, 1986).

Cognitive liberation

Shared grievances, however, are never enough to bring about change on their own. In his three-point "political process model," sociologist Doug McAdam argues that in order for social change to take place aggrieved groups must have a shared sense of hope that change is actually possible; something he refers to as *cognitive liberation* (McAdam, 2010). In this regard, the timing of the zenith of the Civil Rights movement was not coincidental. Notably, African Americans had fought in all four branches of the Armed Forces during World War II, a conflict for which one of the primary justifications for fighting was the inherent racism of Nazi ideology used to justify the extermination of six million Jews and other categories of people deemed inferior. For African American veterans returning to Jim Crow segregation and other forms of overt racial discrimination, the hypocrisy could not have been more jarring. Seen this way, Civil Rights activism was a source of reckoning with the central contradiction of American culture – fighting for freedom abroad while oppressing entire classes of people back home.

Demographic shifts tied to the growth of the US economy by mid-century also increased hope. Despite all the obstacles placed in its way, the African American middle-class had expanded considerably, by mid-century, meaning more than ever before Blacks were able to attain higher levels of education and hold higher status occupations as doctors, lawyers, and college professors. With higher economic status also come a degree of political confidence.

A broader context sometimes neglected in this history is the anti-colonial movements during the first half of the 20th century which culminated in the independence of India and Pakistan from the British empire in 1948. Both Martin Luther King Jr. and Malcom X were clear in the inspiration they derived from these "third world movements" which would place the American Civil Rights Movement in a broader global context. The practice of civil disobedience, for example, was largely inspired through its use by Mahatma Gandhi during the drive for independence.

Political opportunity

A second element of McAdam's political process model is the availability of *political opportunities*. In that regard, the *Brown vs. Board of Education* decision by the Supreme Court in 1954 that deemed public school racial segregation unconstitutional was truly a watershed moment. What began as a set of grievances and a collective sense of rising expectations now had legal teeth. Civil Rights activist could feel the federal government, to a greater degree than before, had their backs. All that was needed now was a corresponding sign from the people that they would commit to a national movement for change. That sign came in the form of a mild-mannered seamstress from Montgomery, Alabama who, through her lone determination stood up to power by refusing to give up her seat in the front of a public bus to a white man. Well, at least most of that previous sentence was true. Rosa Parks was a seamstress from Montgomery, Alabama who made national news through her public transport protest against Jim Crow segregation. She, however, certainly did not work alone, and had long been an active member of the local branch of the National Association for the Advancement of Colored People (NAACP).

Indigenous organization

The point of bringing up Rosa Parks' history of activism and institutional ties to the Civil Rights Movement is not to in any way diminish her courage and determination. Rather, be it in a Hollywood movie or a children's history book, there is a tendency to recount narratives that focus on a single purposeful protagonist who, through hard work and perseverance, overcomes the odds to attain moral victory. Real life is of course much more complicated than that, and sociology reminds us that nothing happens without the coordinated action of many people, what McAdam refers to in his third point as *indigenous organization*. As discussed in the chapter on multiplexity, preachers with congregations, college students, organizations such as the NAACP and the Congress of Racial Equality (CORE), and multiple layers of social networks bolstered social ties and the interpersonal trust necessary to carry out protests and acts of civil disobedience. These socially grounded political actions accumulated and led eventually to permanent changes in federal law. Solidarity built upon a history of interpersonal trust and communal ties is thus an essential element to lasting social change.

Since the 1960s, the Civil Rights Movement has been a source of inspiration for many groups seeking to promote social change, and has served as a kind of template for sociologists to understand other social movements such as LGBTQ rights, and even the environmental movement. Much has changed, however, since the 1960s in both the degree and kind of inequality faced in the United States, and the role electronic media plays in our lives, now versus then. American households saw the violent attacks of police on peaceful Civil Rights protesters on one of three national television networks that existed at the time.

Today, though police violence against non-violent African Americans persists, people have much more choice in the media they can consume and, perhaps more importantly, content providers are more targeted in terms of matching media content with the audience they want. Despite the legislative successes of the Civil Rights Movement, we continue to live in a discriminatory society, punctuated by acts of racial violence, now captured on smart phones and posted on social media. The ability to record heinous behavior as it occurs is greatly facilitated by advanced telecommunication, and yet the ability to ignore it is, as well.

Forecast for change: it's already happening

Within sociology, it is always folly to make claims about future outcomes in human society. There are simply too many moving parts, wheels in motion, and disparate variables at play for us to get a handle on what comes next. We can, however, take what we have learned from the past and use it to formulate a better understanding of the present. In that regard, the present provides us an unprecedented opportunity to grasp social change, namely Covid-19. As

I write this, we are nearing the end of the initial confinement period in the United States for the global pandemic caused by the corona virus. With the exception of a small number of centenarians, no one alive has ever experienced a global crisis quite like this one. As any sociologists of natural disasters could point out, crises have a tendency to bring out the best and worst in the societies affected.

Perhaps the most astonishing thing after it became apparent we were all under a global threat, was how quickly, in a matter of weeks, people across the world accepted measures that would help protect those around them. Small businesses, corporations, schools, universities, and households everywhere made enormous sacrifices, foregoing income, entertainment, job security, and a social life for the greater good of public health, but especially for those most at risk of being harmfully impacted by the effects of the virus. In that sense, you could argue that in all of human history we have never seen a greater expression of global solidarity, a notion captured in the pithy but somewhat accurate phrase, "we're all in this together." That of course begs the question, are we really all in this together?

A clear answer to that question and that of the broader impact of Covid-19 on human society is likely years away and will be debated for a long time to come. From the sociological lens of inequality, however, there are already some early indications of the uneven effect this disease has had on different categories of people in the United States and throughout the world.

The stratified impact of Covid-19

Age

From an epidemiological perspective, the most obvious category to start from is age. Both a blessing and a curse, from early on, it was evident that fatality rates for children infected with Covid-19 were relatively low – less than 0.01 percent for 0- to 9-year olds – while nearly one out of five people 80 years and older with the virus were dying (Berezow, 2020). Sadly, nursing homes were one of the early "super-spreader" sites where relatively large numbers of people were fatally infected (Godfrey, 2020). Beyond the public health implications, as the weeks and months wore on it became apparent that the youth would not be spared. Specifically, young working-age adults between the ages of 16 and 24 were often the first ones to lose their jobs during the national shutdown of restaurants, coffee shops, and hundreds of thousands of retail outlets – precisely those places where people often acquire their first job.

During the early weeks of the crisis, workers in this age range were three times as likely to be laid off or put on furlough as older workers (Rogers, 2020). As May turned into June 2020, it also became clear that traditional fallbacks for summer break – internships, camp counselor, lifeguard at the local pool, volunteer work, world travel – would also be unavailable. For those who did find employment, opportunity was disproportionately found in the "front lines" of grocery stores and home improvement outlets where the average customer faced little risk of infection, but where cashiers, for example, who had multiple interactions with many different people throughout the course of a day faced greater hazard.

Jobs

In the early days of the pandemic, it was clear that access to work was stratified by age, but workers themselves were stratified between "essential workers" who continued to work in place and "remote workers" who could continue to work from home. Much has been made about the awkwardness of home offices wherein online meetings are interrupted by doorbell summons for postal deliveries, and children, pets and spouses inadvertently pass in the background. Also, it can be hard to focus when the immediate demands of domestic life – dirty dishes, dirty clothes, home schooling – surrounds you (Harris, 2020).

That said, the remote worker is likely much better off than the essential worker whose status correlated directly with higher risks of potential exposure to the coronavirus. Healthcare workers are among the highest risk categories of people who face day-to-day interaction with already infected patients. When we look more closely at who occupies these positions, we see disproportionately large numbers of native-born minority groups and immigrants caring for the elderly, changing bedsheets in hospitals, and even among the staff of medical doctors (Kakande, 2020).

Perhaps less obvious but essential nonetheless are the jobs in food processing and the distribution of goods where workers faced risks of infection from Covid-19 equal to or higher than that of healthcare workers. As we learned in Chapter 4, meatpacking has long been one of the most dangerous jobs in America; slashing animal flesh in confined spaces with sharp metal objects was never a recipe for worker safety. In the time of Covid-19, however, the assembly-line nature of abattoir labor has made for a truly hazardous work environment. In the United States, major outbreaks of Covid-19 have occurred in slaughterhouses across the country. Despite the death of some 44 meat processing workers in April 2020 from the coronavirus, President Trump declared meatpacking a "critical infrastructure" that must be maintained for the good of the economy (Scher, 2020). Notably, a similar action was not taken when healthcare workers complained of major shortages of personal protective equipment such as masks and other items essential for their safety.

Another relatively invisible group of workers disproportionately hit by Covid-19 were people essential to the delivery of everything ordered online to our front door. Amazon distribution centers were hit disproportionately by the coronavirus, with the employer being accused of not supplying employees with sufficient personal protective gear to reduce the likelihood of infection (Bellafante, 2020). In all three sectors of employment – healthcare, meat processing, and goods distribution – initial grievances were followed by active protests on the part of workers who attained national attention about their predicament.

In the case of healthcare workers, some parts of the country responded by increasing wages for labor now recognized as truly essential to the nation's fate. In the case of distribution workers, Amazon fired employees who were making efforts to organize a union in April 2020 (Paul, 2020). On May 1, 2020, Amazon, Target, and Walmart employees, among others across the country organized a national "sick-out" day, calling attention to what they felt to be the neglect of essential worker safety at a time when they were most needed. Much of the protests centered around the insufficient provision of protective gear and the lack of transparency regarding the number of workers who had come down with the virus (Beckett, 2020).

Gender

Covid-19 also provides us the opportunity to revisit the sociological theme of intersectionality. The massive layoffs in hospitality, restaurants, and retail, for example, have disproportionately affected women who concentrate in those sectors of the economy; whereas before the pandemic women were slightly more employed in the labor force than men, by April 2020 their rate of unemployment was 3 percent higher than men (Kurtzleben, 2020). This is in direct contrast to the previous economic downturn in 2009 – the "he-cession" recession where the massive downturn in construction meant higher rates of unemployment for men than women. This time around, however, in addition to facing greater unemployment, women also found themselves at home with children, disproportionately shouldering the management and implementation of homeschooling (Harris, 2020). When it comes to house-hold responsibilities and childrearing, long held cultural assumptions about gender appear hard to shake.

Race

The starkest disproportionality in the face of Covid-19, however, has been the higher percentage of cases and death among people of color, African Americans, in particular. The best estimate based on 40 states that collected data on race is that the rate of Covid-related deaths among Blacks is 2.4 times that among whites, and 2.2 times that of Latina/os and Asian Americans (APM, 2020). The statistics for Native Americans is harder to come by because the population itself is small, but in those regions where they concentrate, the data is clear – Covid death rates are eight times greater than whites in New Mexico and five times greater in Arizona, with a particularly high concentration in both states on the Navajo reservation (APM, 2020). From a purely epidemiological perspective, comorbidities can be pointed to as the immediate explanation for these discrepancies. We know that Native Americans and African Americans are two groups with relatively high rates of diabetes, cardiovascular disease, and high blood pressure, all pre-existing conditions that can make succumbing to Covid-19 more likely.

Access to healthcare

By focusing on individual well-being, however, we neglect the sociological context that shapes these outcomes. Worse than that; in concentrating our attention on pre-existing conditions, we risk attributing Covid-related outcomes to the people who have fallen ill themselves; i.e., their bad diets, smoking, lack of regular exercise. As a culture, we are receptive to these kinds of explanations; when we look up solutions for our own ailments, we are often instructed to improve our personal health. This typically necessitates adjusting our diet, getting exercise, and dialing back our use of alcohol, cigarettes, and other things that might threaten our well-being. What we are much less likely to come across is recommendations that we as a society provide broader access to healthcare, with a particular emphasis on historically underserved populations such as African Americans and Native Americans.

How is this relevant to the sociological context of Covid-19? To give one relevant example, testing sites for the virus in Dallas, Texas were found to be easily accessible in relatively wealthy neighborhoods and nearly impossible to access in the poorer parts of town (McMinn, Carlsen, Jaspers, Talbot, & Adeline, 2020). Much of this can be explained by long-term trends in the privatization of healthcare – where people can afford to pay for their own healthcare providers, they have better knowledge of their exposure to the coronavirus than those places that rely on overworked community centers or hospital emergency rooms for primary care. In a market-driven society, this makes good sense – those who can afford it, should be able to get the services they need; a great incentive for all of us to work harder to attain the good life. Perhaps; but as Covid-19 makes overwhelmingly clear, we do not live in a purely market-driven society. When intensive care units in hospital become overwhelmed with patients due to the pandemic, we put the collective well-being of our society at risk.

The uneven impact of American interdependence

As we learned during the global pandemic, "social distancing" and wearing masks, not unlike vaccinating your children for the measles and polio, only really makes sense in the context of the collective good. French sociologist Emile Durkheim would point out that, whether we admit it or not, our default behavior in modern life is to work towards the common good of everyone; what he called organic solidarity. In that regard the United States is not unique – since the establishment of the Americas as a colonial territory, we have been utterly interdependent on each other for our survival. Problems arise, however, when we begin to contrast the official account of this interdependence – *E pluribus unum*; the national motto is right there on all our cash and coins – with our lived experience. The official version is perhaps best represented by the celebration of Thanksgiving. In this annual ritual, people across categories of race, class, immigrant origin, and religion gather in their homes to commemorate an early colonial meal celebrated among European pilgrims and the ancestors of the Wampanoag Nation at Plymouth, Massachusetts in 1621.

Our national union, unfortunately, was more forced and violent than what we learned and relearned every year in elementary school. Just two years before the first Thanksgiving, the first cross-Atlantic ship bearing en-slaved Africans as its cargo arrived on American shores. This set in motion a centuries-long exploitation of African Americans, first as property and the productive base of the agricultural economy, and then as second-class citizens controlled through violence and systematically denied access to capital and political participation in the post-Civil War era (Desmond, 2019).

The plight of indigenous people on American soil was no less grim. After the Revolutionary War and the founding of the United States as an inde-pendent nation, it became clear that the country's territorial expansion would require the forced removal of native people from their lands. This was also done violently, undergirded by the dual ideologies of Manifest Destiny – God's will that the continent be taken over by European-origin settlers – and white supremacy; the notion that whites are self-evidently superior to non-European people (Gilio-Whitaker, 2019).

Armed with these worldviews, those in power could treat non-whites as they saw fit to accomplish both economic goals of capital accumulation, and political goals benefitting elites by ensuring working-class divisions based on race. Sociologist W.E.B. Du Bois captured this last notion best in the 1930s when he recognized that many working-class whites were happy to receive the "psychological wage" of being white and the privileges that entailed in lieu of the economic benefit they might have gained by joining in solidarity with other workers who happened to be Black (Du Bois, 2017).

Divisions along the lines of race persist to this day and an analysis of the disparities in Covid outcomes – the highest rates of cases and Covid-related deaths are in Native American and African American subpopulations –would be incomplete if we ignored the nation's collective history of violence to-wards and neglect of its racial minorities. Speaking to this last point, despite the rhetoric of "We're all in this together," it was remarkable how quickly wealthy residents of major American cities were willing to turn second homes or boats in posh vacation spots into remote work sites, leaving less fortunate city dwellers, including teems of essential workers, to figure things out for themselves (Tully & Stowe, 2020).

Thus, even a "zoonotic virus" crossing over from a wild animal species (likely bats, in this case) must be looked at through a sociological lens to fully understand its impact. African Americans, Native Americans, and poor peo-ple are not genetically pre-disposed to coming down with a virus to which the human species has never before been exposed. A history of exploitation and neglect points to a much more clear-eyed explanation to this particular disparity. Meanwhile, solutions that seek to provide more equally distributed access to healthcare, economic opportunity, and a just treatment before the state could transform the notion, "all in this together," from a falsehood into a commendable collective goal.

George Floyd: the transformative potential of a national tragedy

Unfortunately, we as a nation appear to be a ways from making that particular transformation. Just as the period of national confinement appeared to be loosening up at the end of May 2020, a second contagion began to spread across the country. Ground zero this time was in the Powderhorn community of Minneapolis. After allegedly trying to pass a counterfeit $20 bill at a convenience store to buy cigarettes, George Floyd, a 46-year-old Black man was arrested on the street. Floyd had been a bouncer at a local nightclub, but lost his job as a result of the Covid-19 shutdown. While in police custody, though it is not altogether clear why, Floyd ended up on the pavement on the passenger side of the patrol vehicle across the street from where he was apprehended (Hill et al., 2020).

Though handcuffed and in a prone position, MPD Officer Dereck Chauvin began putting pressure on the back of Floyd's neck with his left knee while two other officers held down Floyd's legs. Chauvin remained in that position for nearly nine-and-a-half minutes, getting up only after emergency medical personnel told him to get off Floyd's motionless body. In the time leading up to that moment, security cameras and two witnesses with smartphones captured the entire incident, including Floyd crying out 16 times, "I can't breathe." Within an hour after being taken away in an ambulance, Floyd was pronounced dead from heart failure. An independent autopsy commissioned by Floyd's family found his death to be a homicide by asphyxiation (Hill et al., 2020).

By itself the violence inflicted on George Floyd was heinous and tragic. His murder, however, was preceded by the shooting death of a Black jogger, Ahmaud Arbery, at the hands of two white men in Brunswick, Georgia, and the wrongful murder of Breonna Taylor, an emergency medical technician, by two white police officers in her own apartment in Louisville, Kentucky. These kinds of incidents did not begin in 2020; the list is long of police and vigilante killings of unarmed African Americans, and every few months another episode reminds us of the violent legacy of racism remaining in our culture since the time of slavery. That said, there was something about the captured image of Dereck Chauvin's knee on George Floyd's dying body and the expression on the police officer's face, hands placed casually in his pockets, which conveyed confidence that his actions were entirely in the right.

For many Americans, this was too much to take sitting down, and major protests erupted across towns and cities in all fifty states the week following Floyd's death. The double blow of a global pandemic threatening public health, and the historic economic slowdown resulting in unemployment near 15 percent, was now met by anti-racist protests and political mobilization reminiscent of the unrest that took place after the assassination of Martin Luther King Jr. in 1968.

At the time of King's death, a sense that Civil Rights legislation had not done enough to address persistent racial disparities, and growing opposition to the Vietnam War resulted in a tumultuous and violent year of protests in American society. In the current day, it appears the intersection of nature and society has brought us to the brink of massive social upheaval, reminding us we are not as separate from the world's ecology as we would like to believe. Covid-19 has laid bare in many ways the underlying forms of inequality that continue to plague our society. Moreover, it underscores intersectionality in the broader sense: George Floyd's Covid-induced unemployment only exacerbated his status as a Black man in a poor neighborhood whose confrontation with the police was not likely to end well. The vast availability of smartphone technology and the omnipresence of security cameras means, increasingly, we are all witnesses to these kinds of atrocities.

Given that fact and the outrage expressed across categories of Americans for George Floyd's murder, we might ask ourselves, what is the likelihood of change this time around? That's not an easy question to answer, but it is becoming clear, as it was to W.E.B. Du Bois a century ago, that no meaningful social change of any kind in this country can take place without first addressing the injustices of race.

In the context of Covid-19, with people working remotely and a relatively high rate of unemployment, one relative advantage for some is a bit more time to read and think about injustice in the world and, just maybe, more time to do something about it. Consider for a moment how much less the turnout for protests incited by the murder of George Floyd would have been had it happened prior to the pandemic. Beyond the sheer numbers of people, another aspect of the protests hard to ignore was the diversity of participants. Whether in Seattle, Minneapolis, or Washington, D.C., anti-racist activists appeared to represent the vast swath of race and ethnic origins that constitute contemporary America.

From a sociological point of view, that is perhaps the greatest source of hope coming out of this bleak episode in our recent history. This expression of solidarity for a common cause among citizens of different backgrounds in cities and towns across country may be the best indicator yet that we as country have the potential to generate greater levels of mutual trust among ourselves. As we will see in the next chapter, social science research tends to backs this up.

Glossary

Cognitive liberation   A shared sense of hope among groups with a shared sense of injustice that change is possible; that dedicating time and energy to a cause is worth the effort.

Indigenous organization   Solidarity built upon a history of interpersonal trust and communal ties; an essential element to building lasting social change.

Political opportunity The moment at which the success of a social movement is most likely, often because of changes in the legal environment
and/or precipitous events which signal a shift in public opinion and an
opening for change.

Psychological wage Term coined by sociologist W.E.B. Du Bois recognizing that many working-class whites in the 1930s were happy to benefit from the mere status of being white, even if they would have much
to gain economically by joining in solidarity with other workers who
happened to be Black.

Zoonotic virus A virus present in the human population that crossed over
from a wild animal species.

Questions

1   Doug McAdam's political process model was inspired primarily by the
Civil Rights Movement. Do a little research on a more contemporary
movement – e.g., Black Lives Matter, Standing Rock opposition to oil
pipeline construction, university divestment from fossil fuels – and map
out as best you can the three elements of cognitive liberation, political
opportunity, and indigenous organization on that particular conflict.
Can your chosen social movement claim success? How do you know?

2   How did COVID-19 impact your life? Do you see the stratified effects
of age, occupation, gender, race, and access to healthcare playing out in
predictable ways in your household? In which ways do you think we
really are all affected by the corona virus equally? How do you think the
perception of COVID-19's unequal impacts on different categories of
people affected the government's response to the pandemic?

References

APM. (2020). *The Color of Coronavirus: COVID-19 deaths by race and ethnicity in the
US*. Retrieved from https://www.apmresearchlab.org/covid/deaths-by-race. Accessed September 16, 2021.

Beckett, L. (2020). Whole foods workers hold 'sick-out' to demand hazard pay during pandemic. *The Guardian*, March 31.

Bellafante, G. (2020). We didn't sign up for this': Amazon workers on the front lines.
New York Times.

Berezow, A. (2020). Coronavirus: COVID deaths in U.S. by age, race. *American
Council on Science and Health*, June 23. Retrieved from https://www.acsh.org/
news/2020/06/23/coronavirus-covid-deaths-us-age-race-14863. Accessed September 16, 2021.

Blau, P. M. (1977). *Inequality and heterogeneity: A primitive theory of social structure* (Vol. 7):
Free Press New York.

Chrisafis, A. (2019). Marseille falls apart: Why is France's second city crumbling?
The Guardian, 6.

Desmond, M. (2019). In order to understand the brutality of American capitalism,
you have to start on the plantation. *The New York Times Magazine: 1619 Project*.

Du Bois, W. E. B. (2017). *Black Reconstruction in America: Toward a history of the part which black folk played in the attempt to reconstruct democracy in America, 1860–1880*: Routledge.

Gilio-Whitaker, D. (2019). *As long as grass grows: The indigenous fight for environmental justice, from colonization to Standing Rock*: Beacon Press.

Godfrey, E. (2020). 'We're literally killing elders now'. *The Atlantic*, April 29. Retrieved from https://www.theatlantic.com/politics/archive/2020/04/coronavirus-especially-deadly-nursing-homes/610855/. Accessed September 16, 2021.

Harris, E. A. (2020). 'It was just too much': How remote learning is breaking parents. *New York Times*, June 12.

Hill, E., Tiefenthäler, A., Triebert, C., Jordan, D., Willis, H., & Stein, R. (2020). 8 Minutes and 46 seconds: How George Floyd was killed in police custody. *The New York Times*.

Kakande, Y. (2020). We are not enemies. We are essential workers. *New York Times*, May 18.

Kurtzleben, D. (2020). Women Bear the Brunt of Coronavirus job losses. *NPR*.

Leras, M. (2020). *Marseille turns Green with election of first woman mayor*: Reuters.

Loewen, J. W. (2018). *Sundown towns: A hidden dimension of American racism*: The New Press.

McAdam, D. (2010). *Political process and the development of black insurgency, 1930–1970*: University of Chicago Press.

McMinn, S., Carlsen, A., Jaspers, B., Talbot, R., & Adeline, S. (2020). In large Texas cities, access to Coronavirus testing may depend on where you live. *NPR*, May, 27.

Morris, A. D. (1986). *The origins of the civil rights movement*: Simon and Schuster.

Paul, K. (2020). Amazon fires two employees who condemned treatment of warehouse workers. *The Guardian*.

Rogers, T. N. (2020). Gen Z is going to get slammed even worse than boomers by coronavirus layoffs. *Business Insider*.

Scher, I. (2020). Almost 12,000 meatpacking and food plant workers have reportedly contracted COVID-19. At least 48 have died. *Business Insider*.

Tully, T., & Stowe, S. (2020). The wealthy flee coronavirus. Vacation towns respond: stay away. *The New York Times*, March, 25.

Wainwright, O. (2013, 2013/04/02/). Sun, sea and super-sizing: Marseille – The port city once notorious for gangs, drugs and violence – is in the grip of a pounds 6bn rebirth. Will its flashy new architecture, including a giant mirror by Norman Foster, make it a worthy Capital of Culture? Oliver Wainwright reports, Article. *The Guardian (London, England)*, p. 16. Retrieved from https://bi-gale-com.ezproxy.uvm.edu/essentials/article/GALE|A324539417?u=vol_b92b

10 Strategies for restructuring an unsustainable society

Social scientist, Thomas Pettigrew spent his youth growing up in the segregated American South. He points to a moment in childhood when he and his nanny went to see a Humphrey Bogart movie as a life-changing event (Taylor, 2015). At the theater, they were told his caretaker could not enter because she was Black. This unequal treatment towards someone he knew and loved enraged Pettigrew. Much later, as a graduate student at Harvard, he began to dedicate his research to the nature of racial prejudice and, more importantly, how it might be eradicated.

At the time he began his studies in social psychology, the dominant view in the field was that people with racial hatred suffered from an "authoritarian personality" (Adorno, Frenkel-Brenswik, Levinson, & Sanford, 2019). The recent trauma of Nazi Germany appeared to reinforce the notion that some people (and entire populations, apparently) were simply predisposed to holding prejudiced and racist worldviews. You could even take a brief quiz to see if you yourself had authoritarian tendencies.

The problem with this approach was it provided practically no room for social change; some people appeared to just be born with an authoritarian disposition. Under the tutelage of Gordon Allport, Pettigrew sought to establish an explanation more clearly grounded in social structure. Decades before, Allport established his *contact hypothesis*, through which he argued that, given the right social context, conflict between two groups could be overcome and prejudice reduced (Allport, Clark, & Pettigrew, 1954). Certain criteria, however, had to be met. These include:

- Both groups should have equal status in education, income, and skills or work experience, if possible
- Both groups should have a common goal or task they could work towards and through which they could pool their efforts
- The intergroup relationship should be characterized by cooperation, not competition
- Some authority over the two groups should support and nurture their interactions

DOI: 10.4324/9781003110668-10

The contact hypothesis has inspired decades of research in sociology and social psychology. In 2006, Pettigrew and his colleague, Linda Tropp, published a meta-analysis of over 700 studies, including lab research, surveys, and quasi-experiments in public housing and schools, using this approach (Pettigrew & Tropp, 2006). The central finding: 94 percent of studies in this sample set found a negative relationship between intergroup contact and level of intergroup prejudice, strong empirically-based support of Allport's theory. Moreover, Pettigrew and Tropp found that only 19 percent of these supporting studies followed strictly Allport's criteria, suggesting that intergroup contact, even without the above list of conditions, would likely reduce intergroup prejudice.

If intergroup contact appears clearly in the data as a source of prejudice reduction, what are the negative forces keeping prejudice in place? As we mentioned in an earlier chapter on Race and Justice, structural racism – i.e., the accumulated history of exploitation, discrimination, and exclusion from economic opportunity of Blacks and other racial minorities – is the long view on how racist beliefs and actions persist and are woven into the fabric of American culture and social institutions (Bonilla-Silva, 1997). Straddling the disciplinary boundaries of sociology and social psychology, Pettigrew wanted to know more specifically the experiential origins of prejudice in everyday life. Though many lab-based studies and small-scale research supported the contact hypothesis, it was less clear how well it applied in the real world. To what degree could the criteria established by Allport be relaxed in a context where the nature of structural racism means two groups of different backgrounds are unlikely to meet in a context of equal social status?

The perception of threat as the basis for prejudice

To better understand how racist views and prejudice are perpetuated, Pettigrew and his colleagues propose a force counter to intergroup contact which favors higher levels of prejudice – threat (Pettigrew, Wagner, & Christ, 2010). Their research in this area is inspired by more recent survey work in Germany, a country that has seen high levels of immigration in recent decades. The strange thing here is, in eastern Germany anti-immigrant sentiment is much greater than it is in western Germany which has a much larger proportion of immigrants in its population. Germany, by the way, is not alone in this regard – in France, support for the far-right, anti-immigrant political party comes disproportionately from rural areas where the ratio of immigrants to non-immigrants is much lower than it is in the cities (Chrisafis, 2017). The United States shows similar patterns – cities with the highest concentrations of immigrants have the highest tolerance of immigrants and areas with low levels of diversity tend to be more nationalistic in their views (Sacchetti & Guskin, 2017).

This has interesting implications for strategies to reduce prejudice. The key distinction to be made here is this – threat is perceptual, while intergroup contact is experiential. According to Pettigrew and his colleagues, in the political realm this means that threat can be easily manipulated, especially in the context of little intergroup contact. As mentioned in the previous chapter on risk and trust, US politicians have effectively tied the labels of terrorist and criminal to immigrants, despite the fact that domestic acts of terror including mass shootings and bombings are more likely to by to be carried out by American citizens, and that levels of criminality are much lower among immigrants than non-immigrants.

Put simply, threat is a political tool used by elites to foster divisions within a diverse society for political gain. The tonic for this ailment is intergroup contact. This works because, given the social side of our nature, regular interaction with other people different from ourselves tends to reduce our anxiety about interacting with them. Over time, uncertainty fades with mutual understanding and even friendship taking its place.

By contrast, not interacting with outsiders increases intergroup anxiety. Segregation, either *de jure* and supported by law as was the case before the Civil Rights era, or *de facto*, reinforced by structural racism as it is now in communities, schools, and workplaces across the country, is a horrible recipe for increasing the perception of threat and reducing the experience of interpersonal contact among people of different backgrounds. In identifying racism as *structural* we are acknowledging that shifting our culture away from racism cannot be done simply with a change in attitude. What is required instead is a profound change in social structure. Thankfully, there is some evidence that this is possible.

Structuring integration: Affirmative Action laws

Concurrently with Civil Rights legislation passed by Congress in the 1960s, and continuing into the Nixon administration of the 1970s, a series of executive orders from the White House put in place a set of policies collectively known as Affirmative Action. Though the Civil Rights Act of 1964 made it unconstitutional to discriminate on the basis of race in the hiring of jobs, it did not compel employers to actively recruit race and ethnic minorities and women.

Affirmative Action policy, by contrast, did just that, applying to three major classes of employers; the government itself, firms who received revenue in the form of government contracts, and firms that employed 50 people or more. Affirmative Action laws have been criticized for, among other things, disproportionately benefitting the social mobility of white women relative to other disadvantaged categories, favoring members of minority groups who already come from middle-class backgrounds, and not doing enough to address disparities on the basis of social class. Despite these criticisms, in those sectors of the workforce where it was effectively implemented, Affirmative Action has generated increased economic opportunity for African Americans and other minorities, and has increased intergroup contact.

Perhaps the most telling example in this regard has been the US military. During World War II, troops were segregated by race – an appalling juxtaposition given the anti-fascist/anti-racist rhetoric used to justify US participation in the war effort. As is well established, Hitler himself pointed to the United States as an exemplar in how a racial hierarchy could be effectively managed (Ross, 2018). Since President Harry Truman integrated the armed forces in 1948, however, the US military has become the most racially integrated institution in American society. Currently, Blacks make up of 17 percent of active military, compared to their 13 percent share of the US population (Parker, Cilluffo, & Stepler, 2017).

Of course, as economic opportunity has become more stratified in the United States since the 1970s, military service appears as one of the more viable and secure options available among the working class where race and ethnic minorities tend to concentrate. Sadly, recent years have seen a spike in reports of racism and white nationalism within military ranks (Cooper, 2020). Nonetheless, institutional changes have led to a deliberately more integrated and diverse population of infantry than there would have otherwise been. Leadership matters, and blocking out entire classes of people on the basis of race, gender, or other categories from opportunity makes little sense from the perspective of either social justice or the sound bureaucratic management of institutions, public or private.

Along with lifting barriers to opportunity, sociologists such as Peter Blau, mentioned in the previous chapter, have long recognized that lowering levels of inequality also works towards reducing class-based social distance and better integrating society. Unfortunately, in the United States, policy alternatives such as raising the federal minimum wage and capping executive pay have met with at least as much resistance as have moves towards greater racial and ethnic integration.

Structuring sustainability: environmental legislation

The 1970s, a decade which weathered both the political turmoil of the Vietnam War and the shameful resignation of President Richard Nixon, saw long-lasting structural change in other parts of American life besides workplace Affirmative Action laws. Notably, the environmental movement attained its greatest achievements during this ten-year period, much of which were signed into law by Nixon himself. Let's be clear: Nixon broke the law by ordering a break-in to Democratic Party headquarters at the Watergate Hotel in Washington, D.C., and only escaped jail time thanks to the presidential pardon of his former vice president Gerald Ford. Before that, his escalation of the Vietnam War led to mass demonstrations, one of which resulted in the killing by the National Guard of four protesters at Kent State in 1970. There are many reasons to sneer at Nixon's legacy. Out of dark periods, however, sometimes arise unexpected accomplishments.

Here is a brief list of environmental legislations passed and organizations created prior to Nixon's resignation in 1974 (Climate Central, 2012):

- *The National Environmental Policy Act* (EPA, 1969), requires all federal agencies to produce environmental impact statements stating the possible negative effects federal regulations may have on the environment
- *The Environmental Protection Agency* (1970), conducts environmental assessment, research, and education, and enforces federal environmental standards
- *The National Oceanic and Atmospheric Administration* (NOAA, 1970), A scientific body that monitors the oceans, waterways, and atmosphere to better protect life and property from natural hazards, and better manage the use of marine resources
- *The Clean Air Act* (1970), designed specifically to improve air quality by limiting emissions from both transportation and industrial sources
- *The Clean Water Act* (1972), designed to restore and maintain the quality and biology of the nation's water resources
- *The Endangered Species Act* (1973), designed to protect endangered species and their corresponding natural habitats to the extent that such protection will no longer be necessary

There are, of course, more, and less cynical ways of considering this legacy. At a time when he was sending off young people to fight and die in an unpopular war in Southeast Asia, Nixon enacted some of the most progressive laws since the New Deal legislation of the 1930s. His executive orders that put into place long-standing Affirmative Action and environmental laws, you might argue, were Machiavellian; shrewd and calculated political maneuvers by "Tricky Dick" intended to distract the public from his otherwise bad behavior. You might also argue, so what? Political opportunity may arise when you least expect it; social movements must thus be on the ready to seize these opportunities before they slip away. Occasionally, bad leadership can lead to the unintended consequence of social upheaval, mass demonstrations, political organization among non-elites, and an avalanche of social change.

The mismatch between social structure and socioecological reality

One influential take on unintended consequences comes from a 1936 article on "purposive action," by sociologist Robert Merton. There, he argued the most deleterious aspects of modern life – war, environmental devastation, poverty – are a result of a mismatch between, on the one hand, the ideals and goals of the social systems we have built on the base of a common culture and, on the other, reality.

It is thus not surprising to see that our purposive actions as a society, including the unregulated extraction of natural resources, the militarization of police forces across the nation, the deregulation of banking and finance, and the mass availability of guns, will result in the unintended consequences of mass species die-off, police brutality and mass incarceration, bank fraud and predatory lending, and high levels of suicide among middle-aged men. Though the goals of the original actions – often tied to the generation of greater profit – may have been met, the side effects on the environment and society are devastating.

What if this worked in the other direction, though? What if the "intentions" of reality and the natural world resulted in the unintended consequence of us humans changing social structure? To step back for a bit, sociologists have long talked about the distinction between "structure" and "agency" when trying to accurately account for social dynamics in modern life. Structure is the context into which we are all born; this can refer literally to physical structures such as buildings, houses, and the infrastructure of electricity, water, sewer systems, and transportation that make modern life possible. Structure can also refer to social institutions; i.e., schools, universities, hospitals, corporations, and perhaps most abstractly, the economy. We could also include here pre-existing, historically-based social categories of race, gender, religion, and nationhood that can rigidly frame the canvas of our lives. Agency, by contrast, is what we choose to do as individuals and collectivities within the constraints set for us by social structure.

Nature as structure

A major challenge for environmental sociologists has been to figure out where the natural world fits into this grand scheme of structure and agency. The dominant tendency in the discipline of sociology has been just to ignore nature altogether and focus our energy on the topics we know best – society and its socially derived institutions. The second major approach since the emergence of environmental sociology in the 1970s is to view the environment as natural resources, typically unaccounted for in the capitalist ledger of material inputs for the production of goods and services. In economics, you could also include here the broad category of *negative externalities*, i.e., the negative impacts economic transactions have on people who had no say in a transaction occurring in the first place.

Within environmental sociology the specific use of nature for depositing human-produced waste in the environment is known as a "sink" (Schnaiberg, 1980). Air and water pollution from industry and agriculture are classic examples of this and also, by the way, the *raison d'etre* for legislation such as the Clean Air and Clean Water Acts passed during the Nixon administration. Without regulation, profit-seeking industries will be strongly tempted to reduce costs by dumping waste in the environment. As made evident through the example of VW in the chapter on rationalization, even regulated industries may be tempted to slough off costs on the rest of us by polluting common resources.

Though perhaps obvious to us now, this resource and sink approach to understanding the nexus of environment and society was a major leap forward in at least acknowledging the fact that modern society was utterly dependent on natural resources for its existence. This approach fits nicely into the structure and agency model for understanding social phenomenon by placing nature firmly on the side of structure. Not unlike libraries and political systems, nature existed before we were born, it is hard to change, but with enough agency and collective will, we can shift its structure over time.

Agency in nature

A perspective that is still very much in the minority within sociology, but which is starting to gain traction as viable among critical environmental justice scholars, is the notion that nature itself has agency in the face of structure and existential threat created by the human species. As sociologist David Pellow observed,

> Worms, viruses, ants, water, rocks, mountains, fish, elephants, krill, air/ wind, and trees are just some of the infinite nonhuman beings and things that are affected by environmental injustice but that also exert their own influence on the character of those conflicts in particular and on the course and trajectory of human society and ecology more generally.
>
> (Pellow, 2017, p. 19)

There are still those within mainstream sociology who would wince at the suggestion that animals, plants, and the Earth itself could have anything approaching what we consider agency. What Pellow seems to be suggesting we do, though, in addition to adopting a greater sense of humility, is to extend our sense of social interdependence to that of the "more-than-human" world. Taking his cue from the Black Lives Matter movement, Pellow underscores the *indispensability* of all living beings and categories. In direct opposition to white supremacy and human dominionism, we need to understand, as Martin Luther King Jr. wrote, that "Injustice anywhere is injustice everywhere…. In a real sense all life is inter-related. All [people] are caught in an inescapable network of mutuality" (as cited in Pellow, 2017, p. 28).

Though we collectively come from a diverse set of backgrounds, the common experience of oppression by those with greater power than ourselves brings us together across categories of race, class, gender, and even species. Including nature in the list of willful agents subject to exploitation and yet capable of resistance and liberation, changes the calculus of how socioecological change might play out in some important ways.

First, let's be clear: We are still using human language, ideas, and institutions to make theoretical space for the inclusion of more-than-human agents in what is clearly the human realm of culture and politics. Grizzly bears will not be running for office and chickens will not be taking to the streets to protest poultry oppression. Animal liberation and environmental protection, for that matter, are still and will remain for the foreseeable future the political prerogatives of human beings. That said, indispensability requires that we re-center our place in the socioecological cosmos. But how do we do that? How do we step outside ourselves? How do I remove myself from a comfortable existence in which so long as I do not personally commit racist acts or defile nature directly, allowing supply chain intermediaries to buffer me from animal slaughter, pipeline construction across indigenous lands, uranium mining, child labor, and deforestation of the Amazon, I feel fine?

When time stood still and, then, it didn't

The flourishing of the Black Lives Matter movement during the global spread of Covid-19 – the most consequential intersection of nature and society in our lifetime – can give us some insights here. Think for a moment how the perception of time itself became distorted during the pandemic. In mid-March 2020, I wished my students at the University of Vermont a relaxing spring break, knowing they would be back for the second half of the semester in about 10 days. In that period between two weekends, however, the world turned upside down. All classes had to transition to online remote teaching; my two kids, then eight and ten, would have to be home schooled; both the NBA season and collegiate basketball's March Madness competitions were canceled in a matter of days; cruise ships were stuck in port with passengers quarantined on board; and over 90 percent of the world's commercial airlines fleet was grounded. Abruptly, time itself seemed to stop.

Most of us could not leave home and those who could witnessed scenes akin to a science fiction movie; major world cities without any moving cars, no one on the streets. In more suburban areas with less population density, foxes, woodpeckers, bears, moose, and other species seemed to roam and rush more freely through backyards and overhead. Was there really more wildlife around, or did we just finally have the time to pay attention? The skies themselves, less cluttered with air traffic, appeared bluer during the day and clearer at night; the planet Venus was bright enough to cast shadows through my upstairs window in the late evening, a fact facilitated by its coincidentally close approach to Earth at the time, I found out later. The important thing is I noticed.

As the pace of life slowed down, many of us, it is true, streamed more movies and entertainment online, but many of us also began to stream outside, walking, running, and biking in our neighborhoods. In doing so, we noticed people and things nearby we had never taken the time to pre-Covid. This slowdown was also reflected in our energy consumption. The price of gasoline fell dramatically as many people were not working or, if they were, did so remotely, driving down demand for transportation (Brower & Sheppard, 2020). Worldwide, carbon emissions dropped by 17 percent as the coronavirus accomplished something decades of international conferences and accords on climate change and the International Panel on Climate Change could not (Borunda, 2020).

Then, after two months of confinement, we were jolted back into history. With the murder of George Floyd, it suddenly seemed anything was possible. After a decade of activism that received lukewarm acknowledgement from the general public, at best, and, at worst, out and out derision, the Black Lives Matter movement blazed across the consciousness of Americans and people the world over. After two weeks of mass protests and confrontations, the city of Minneapolis announced it would defund the police department; a few municipalities across the country claimed to follow suit, many more began openly talking about this possibility among others to rethink community policing.

Beyond the immediate question of police brutality, in cities and towns across the South, monuments to Confederate war "heroes" were taken down out of recognition that these formerly celebrated individuals fought for the enslavement, torture, and dehumanization of Black Americans. At NASCAR racing events, long tolerant of overt displays of white supremacy, the Confederate flag was officially banned. The governor of Kentucky announced healthcare for African Americans would be free for all who need it.

Protests across the country appeared to be working; non-institutional politics – outside the voting booth, state capitols, and offices of elected officials – were alive and well after what seemed like a lifetime of absence. Coming back to the question of social change, why this time did so many people join together in solidarity around a movement that had been voicing the same grievances for much of the early 21st century?

As discussed in the previous chapter, the Civil Rights Movement taught us that grievances are never enough for a movement to gain broader traction. In addition, the push towards social change requires hope, rising expectations, precipitating events and movement resources, including network ties that bring people together for a common cause. The question of social ties appears particularly germane to the Black Lives Matter protests during the pandemic, given the quantity and diversity of people who participated therein.

Why people get involved with movements for social change

You may recall from Chapter 4, sociologist Doug McAdam's research (1986) on why middle-class students from outside the South were motivated to be part of the Freedom Summer campaign of 1964. The theme there being, it is one thing to say you are against racism and discrimination, but it is a whole other level of commitment to actually show up and do something about it. This last point was all the more significant given the level of violence inflicted by white supremacists at the time, including a series of murders of Civil Rights activists.

Who, coming from a relatively secure white, middle-class background, would choose to participate in this movement? Along with supporting the ideals of the movement and being "biographically available," a key characteristic of Freedom Summer recruits who showed up was that they knew someone else who was participating. Much social movement research since that time has backed this up – the greatest predictor that you will attend a political protest or be involved in political organizing is that you already know someone involved in these activities.

As a general rule, people like to know they are not alone in supporting a cause. Thus, one of the central ironies of the Covid experience has been, at a time when many of us were socially isolated, people managed to find connection to others through the grief and anger surrounding the suffering and death of one man representing centuries of oppression; victim to the malicious, racist underbelly of American culture.

Current prospects for socioecological change

Recently, McAdam has begun to apply his social movement perspective to climate change activism in the United States. Here, he sees some hope, especially in the growing number of young people who have become part of the movement (McAdam, 2017). College students, in particular, in part because they will have to live with the long-term effects of climate change, but also because of their relative biographic availability, will likely play a crucial role in climate action. A wave of successful fossil fuel divestment campaigns on college campuses across the country in recent years may be an early indication of this trend.

On the whole, however, McAdam is reserved about the prospects of climate change activism ever taking off as a transformative movement, supported by large portions of the population. A big challenge in this regard has to do with the framing of the climate change movement around a specific identity. No one "owns" climate change as a political identity the same way African Americans, for example, own Black Lives Matter. If anything, McAdam can point to research which finds conservatives, including "climate deniers," to be the ones who feel most strongly about climate change in that they are adamantly opposed to policies that would mitigate its effects.

Not all sociologists share McAdam's skepticism, however. For John Bellamy Foster (2019), the year the global climate movement finally "caught fire" was 2018. In August that year, Greta Thunberg began her school strike outside the Swedish Parliament, demanding policymakers finally take seriously the looming threat of environmental catastrophe posed by climate change. The fact she was 15-years-old at the time only underscored her message that adults were shirking their responsibilities towards future generations and, indeed, their own children by neglecting to address the deleterious impact fossil fuel consumption is having on the planet. Other young people noticed and thousands of similar strikes took place the world over, most notably on March 15, 2019.

Meanwhile on opposite sides of the Atlantic, two national-level movements emerged at about the same time. In the UK, Extinction Rebellion began a series of mass protests, including the occupation of Greenpeace's UK headquarters in October 2018 and, a month later, the blockage of five bridges over the Thames River, bringing to a standstill traffic in downtown London. Mass civil disobedience was now being used as an effective way of calling attention to the urgent nature of the climate crisis.

In the United States, the Sunrise Movement has focused its efforts on influencing lawmakers to act to combat climate change; specifically, it pushed forward the Green New Deal, a congressional resolution that could synergistically address climate, social inequality, and access to jobs while laying out a path of societal transition away from fossil fuel dependence. Sunrise is also a youth-led movement that has built momentum since 2018; it endeavors to bring about policy-based social change through institutional means, specifically, the US Congress.

Radical, interstitial, and symbiotic transformations

Institutional politics is but one potential (and challenging) path to socioecological movement success. The late sociologist, Erik Olin Wright (2010) argued that social transformation tends to take three primary forms; rupture, interstitial, and symbiotic. *Rupture* implies that before we build anything up, we have to tear everything down; the definition of radical social change, evident in the removal of the aristocracy from power during the French Revolution, and the violent upheavals that took place during the 20th-century Chinese and Russian communist revolutions. Though worth considering in a broader conversation about societal transformation, Wright is doubtful these kinds of violent ruptures under current conditions could lead to a satisfying outcome, especially if the majority of the population is not in favor of a radical break with contemporary life and societal expectations about the future.

Interstitial transformation, in Wright's view, reflects efforts at the individual- and community-level to operate outside global capitalism; finding space between the cracks and fissures of society's dominant institutions to establish alternative ways of living and relating to one another with the goal in mind of long-term social transformation. Transition communities, Slow Growth and the Degrowth movement – at the moment, all more evident in Europe than the United States – are examples of interstitial efforts to bring about socioecological change by working on the margins of the dominant capitalist system. In these in-between spaces, economic growth is not an imperative, and energy alternatives to fossil fuels are prioritized. As Wright argues, if the appropriate metaphor for ruptural transformation is war, the appropriate metaphor for interstitial change is ecological competition. In a world of increasingly constrained natural resources, for example, conservation and the communal management of common resources could prove to be highly adaptive strategies.

Symbiotic transformation involves working within established institutions to bring about structural change. The Sunshine Movement's advocacy of the Green New Deal cited above is a good example of this, as were 20th-century Social Democratic governments in Europe, wherein the working class, far from being a radical revolutionary class as envisioned by Karl Marx, was a key player in institutional politics, represented directly by its own political party. This strategy, as Wright points out, involves the most concessions to the dominant capitalist class which resists regulation and restrictions on profit-making. Symbiotic strategies have, however, brought about lasting institutional and social change for the collective good, as was the case for the original New Deal during the Great Depression, Civil Rights legislation passed in the 1960s, and the suite of environmental acts enacted by Congress during the 1970s.

In the next and final chapter, we will consider some important ways sociology can inform your chosen path for bringing about lasting socioecological change. Before we finish this chapter, however, I would like to convey this essential message: prior to advocating for radical, interstitial, or symbiotic socioecological transformations, the most important first step you can make to assure a more just, sustainable future is to get out in the world to meet and work with other like-minded people towards a common cause. Anything less is just brooding.

Glossary

Agency   What we choose to do as individuals and collectivities within the constraints set for us by social structure; the ability to take action.

Authoritarian personality   A social psychological theory from the World War II era which posited some people were more predisposed than others to holding prejudiced and racist worldviews.

Contact hypothesis   Theory which argues that, given the right social context, conflict between two groups can be overcome and prejudice reduced, provided certain criteria are met, including   Both groups should have similar a socioeconomic status; both groups should have a common objective; their relationship is characterized by cooperation, not competition; and some authority over the two groups supports and nurtures their interactions.

Indispensability   The inherent value of all living beings and categories of things in nature; an argument from a critical environmental justice perspective for the revalorization of human diversity and nature in the face of looming existential threat.

Interstitial transformation   Efforts at the individual and community-level to find space between the cracks and fissures of society's dominant institutions; working to establish alternative ways of living and relating to one another with the goal in mind of long-term social transformation.

Negative externalities The adverse impacts economic transactions have on people who have had no say in whether a transaction should occur in the first place.

Purposive action The deliberate actions we take to sustain a way of life based on our presumed common goals and ideals, but which often result in unintended consequences that threaten societal stability.

Rupture A radical form of social transformation which implies that before we build anything up, we have to tear everything down.

Structure The context into which we are all born; this can refer literally to physical structures as well as social institutions, and can include pre-existing, historically-based social categories of race, gender, religion, and nationhood that directly shape our lives.

Structural racism The accumulated history of exploitation, discrimination, and exclusion from economic opportunity of Blacks and other racial minorities; an account of how racist beliefs and actions persist and are woven into the fabric of American culture and social institutions.

Symbiotic transformation Working within established political institutions and government to bring about social change.

Threat The perception of risk which may be associated with prejudice against subordinate groups in society and manipulated by political leaders to foster divisions within a diverse society for political gain.

Questions

1   What is your own experience with the contact hypothesis? Have you found that spending time in diverse settings has increased your ability to trust and empathize with others? Do you see any holes in Thomas Pettigrew's arguments for greater integration? What would it take for the United States to be open to spending time with a greater diversity of acquaintances and friends? In the current climate, do you perceive that we are becoming more or less integrated as a society?

2   What social movement strategy for socioecological change do you think will be most effective over your lifetime; utilizing rupture, interstitial, or symbiotic tactics? Why would some of the strategies have less of a chance for success?

References

Adorno, T., Frenkel-Brenswik, E., Levinson, D. J., & Sanford, R. N. (2019). *The authoritarian personality*: Verso Books.

Allport, G. W., Clark, K., & Pettigrew, T. (1954). *The nature of prejudice*: Addison-Wesley.

Bonilla-Silva, E. (1997). Rethinking racism: Toward a structural interpretation. *American Sociological Review, 62*, 465–480.

Borunda, A. (2020). Plunge in carbon emissions from lockdowns will not slow climate change. *National Geographic*. Retrieved from https://www.nationalgeographic.com/science/article/plunge-in-carbon-emissions-lockdowns-will-not-slow-climate-change. Accessed September 17, 2021.

Brower, D., & Sheppard, D. (2020). Global petrol demand to plunge as US stops driving. *Financial Times*. Retrieved from https://www.ft.com/content/e87d9f10-cb16-42af-a4dc-6f8ffc725048. Accessed September 17, 2021.

Chrisafis, A. (2017). The real misery is in the countryside: Support for Le Pen surges in rural France. *The Guardian*.

Climate Central. (2012). Lost in Watergate's Wake: Nixon's Green Legacy. Retrieved from https://www.climatecentral.org/blogs/richard-nixon-the-environmentalist-resigned-38-years-ago-today-14776

Cooper, H. (2020). African-Americans are highly visible in the military, but Almost Invisible at the Top. *New York Times*, May 25, 2020.

Foster, J. B. (2019). On Fire This Time. *Monthly Review (New York. 1949)*, 71(6), 1–17. doi:10.14452/MR-071-06-2019-10_1

McAdam, D. (1986). Recruitment to high-risk activism: The case of freedom summer. *American Journal of Sociology*, 92(1), 64–90.

McAdam, D. (2017). Social movement theory and the prospects for climate change activism in the United States. *Annual Review of Political Science*, 20, 189–208.

Merton, R. K. (1936). The unanticipated consequences of purposive social action. *American Sociological Review*, 1(6), 894–904.

Parker, K., Cilluffo, A., & Stepler, R. (2017). facts about the US military and its changing demographics. *Pew Research Center*.

Pellow, D. N. (2017). *What is critical environmental justice?*: John Wiley & Sons.

Pettigrew, T. F., & Tropp, L. R. (2006). A meta-analytic test of intergroup contact theory. *Journal of Personality and Social Psychology*, 90(5), 751.

Pettigrew, T. F., Wagner, U., & Christ, O. (2010). Population ratios and prejudice: Modelling both contact and threat effects. *Journal of Ethnic and Migration Studies*, 36(4), 635–650. doi:10.1080/13691830903516034

Ross, A. (2018). How American Racism Influenced Hitler. *The New Yorker*.

Sacchetti, M., & Guskin, E. (2017). In rural America, fewer immigrants and less tolerance. *The Washington Post*, June, 17.

Schnaiberg, A. (1980). *The environment: From surplus to scarcity*: Oxford University Press.

Taylor, M. C. (2015). *Introduction of Thomas F. Pettigrew: 2014 recipient of the Cooley-Mead award*: SAGE Publications.

Wright, E. O. (2010). *Envisioning real utopias* (Vol. 98): Verso.

11 Socioecological solutions from the inside out

The environmental movement could use a lot more sociology. There is greater openness about its shortcomings than there was 20, 50, and certainly 100 years ago, but it is worth emphasizing just how exclusionary, elitist, and even racist American environmentalism was in its origins. In 2020, amid the street protests and uprisings that took place after the murder of George Floyd, the Sierra Club, one of the largest and most influential environmental organizations in the world, finally apologized for the derogatory views of its founder, John Muir, toward Blacks and American Indians (Melley, 2020). Muir, sometimes referred to as the American "father of our national parks," has long been admired for his trailblazing efforts at conservation and his ability to shape natural resource policy through his relationship with power elites such as President Theodore Roosevelt and Manhattan lawyer/zoologist, Madison Grant (Cagle, 2019).

Himself a Scottish immigrant, Muir was also a white supremacist who kept company with fellow Sierra Club members such as Joseph LeConte and David Starr Jordan, eugenicists who favored the forced sterilization of minority groups to help ensure the genetic dominance of the white race (Melley, 2020). They were by no means alone: Grant, along with being a fellow conservationist, wrote in 1916 one of the most influential tomes in white supremacist "literature" – *The Passing of the Great White Race* – which influenced among other wilderness loving eugenicists, Adolf Hitler (Purdy, 2015).

That was a long time ago, and much has changed, but the exclusionary and elitist origins of American environmentalism linger. Rates of attendance at national parks are still much higher among whites than people of color (Krymkowski, Manning, & Valliere, 2014), and membership in environmental organizations is overwhelmingly white, this despite the fact that numerous national studies find greater sensitivity to environmental issues among people of color than the majority population (Head, Klocker, & Aguirre-Bielschowsky, 2019; Macias, 2016a, 2016b; Taylor, 2014a, 2014b).

Even within the American Sociological Association, the Section on Environmental Sociology has the lowest percentage of non-white membership relative to all other subsections in this relatively diverse academic professional organization (Mascarenhas, Carrera, Richter, & Wilder, 2017). In

DOI: 10.4324/9781003110668-11

government, sociologist Jill Harrison found through extensive interviews and participant observations conducted before the 2016 presidential election that decision-makers within the Environmental Protection Agency did not take seriously matters of social equity and disproportionate impacts on communities of color, leaving lower level advocates for environmental justice struggling to reform the agency from the inside out (Harrison, 2019).

All this suggests that an environmental movement which excludes the youngest, most rapidly growing, and diverse portion of the population is really no movement at all. A progressive shift towards a more just, equitable, and sustainable planet, by contrast, will require a resolute openness to the diversity of human experience. In that regard, sociology can help. I thus conclude this book with five key sociological concepts that can reorient our thinking and actions towards a more inclusive socioecological future.

Work on your diverse multiplex ties

Sociologists studying social networks observe that we are connected to each other in less complex ways than in the past; increasingly, we suffer from social isolation (McPherson, Smith-Lovin, & Brashears, 2006). Moreover, the people we spend time with tend to be an awful lot like ourselves (McPherson, Smith-Lovin, & Cook, 2001). Socioecological change requires we get out of this rut. It is good to hang out with your longtime friends, but it is also healthy for you and the rest of society to spend time with people different from yourself. One of the best ways to do this is through volunteerism. The thing about volunteer work is it almost guarantees you will meet someone different from yourself in significant ways. Though power can be used interpersonally to gain competitive advantage over adversaries, it can also be used to help those in need.

Think of a cause, an issue, something you or your family have been affected by personally, perhaps, or an issue that has really gotten under your skin as of late, and look for ways you can make a difference through your actions. This need not involve protesting in the streets or running for public office (though, that would be fine, too). This should involve, however, regular interaction with people you would not normally encounter. Tutoring, grocery shopping for the elderly, organizing a dinner for new Americans at your place of worship, or more formal volunteer work with organizations such as Teach for America and Habitat for Humanity, all take your relative life advantages and redistribute them for the greater good.

You will not as an individual change the world, but through volunteerism you will generate good will and make society just a little bit more integrated than it was before. The more integrated we become and trusting of people across race, class, and gender backgrounds, the better prepared we will be to face collectively the economic, ecological, and political crises coming our way.

If you are inclined to be more politically active, think about organizations and causes that bridge across categories of class, race, and gender. Multiplexity is particularly powerful when made up of diverse ties, generating empathy and good will among people of different cultural, socioeconomic, and political backgrounds. In 2016, the protests around the placement of the Dakota Access Pipeline near the Standing Rock Sioux Indian Reservation brought in protesters from across the country (including some of my own students at the University of Vermont) in opposition to the pipeline. These actions, which included confrontations between activists and US National Guard troops, brought national attention to an oil infrastructure plan which threatened to contaminate regional groundwater. Unable to ignore the demonstrations at Standing Rock, the Obama administration put a hold on pipeline construction until a thorough environmental review could be carried out (Gilio-Whitaker, 2019). Within a month of entering office, however, President Donald Trump lifted this hold and put through an expedited review, allowing oil to begin flowing through the pipeline in June 2017.

From the outside, it might have appeared that all hope was lost to somehow revive opposition to this now permanent installation of pipeline infrastructure. Activist networks, however, were alive and well, and financial support for legal groups such as Earth Justice permitted challenges in the courts to continue after Trump's edict. Finally, in July 2020, a federal court demanded that the Dakota Access Pipeline be shut down and emptied of its contents until a thorough, years-long environmental impact assessment could be carried out (Eaton & Matthews, 2020). Political linkages outside the Sioux Reservation and across the country made this environmental success possible. The 2016 Standing Rock protests were really just the most visible side of years, generations even, of political organization and planning for indigenous environmental justice.

Multiplex ties, however, are like that – building slowly overtime, rooted in regular interaction and planning with others. In that way, activist networks work within a context similar to what paleontologists Stephen J. Gould and Niles Eldredge (1977) referred to as "punctuated equilibrium" in the evolution of species. That is, the fossil record suggests species can go thousands even millions of years without any remarkable changes in their morphology. Then, suddenly, with a major shift in habitat – climate volatility, an earthquake, or meteor – everything changes and new ways of being emerge. Similarly, with social change, things remain the same, nothing happens, then – bam! – anything seems possible and the old rules no longer apply. The take-home sociological message here – invest in your diverse multiplex ties now for the lasting change you envision in the future. When opportunity finally arrives, you will be ready.

Seek out alternatives to rationalization

Within capitalism, cost reductions in production and profit maximization take precedence. Even in the realm of sustainability, more efficient, resource intensive technologies – e.g., the electric car – are given much more attention than conservation − e.g., driving less. The attractive thing about efficient technologies from the purview of rationalization is, (a) they can reduce labor costs in production and, (b) they do not require consumers to cut back on spending − a disaster from a profit maximization perspective. A socioecological alternative to rationalization would place other considerations above profit. For example, human dignity; the collective health of the planet; the well-being of future generations; freedom from racism and sexism; and a fair collectively agreed upon distribution of opportunities and life chances among human beings.

Covid-19 has been a tragedy for the millions of people around the world who have suffered, died, or lost loved ones to this disease. It has also highlighted, however, the fact that, as a global society, we can and do value other things besides profit. Moreover, though the economic trade-off is significant, we can change our behavior at the global scale. We can survive, for example, and drive less. International commerce and academic research do not require that people fly around the world making handshake deals and attending conferences at fancy hotels in hip metropolitan locales. We can use less toilet paper. Instead of producing more things at lower costs while consumers spend beyond their means, the non-rationalized world shows us that, when required to do so, producers rein in production and consumers cut back on spending. What more, environmental impacts are reduced along the way. Beyond the notable reduction in carbon emissions, it is estimated that even roadkill was reduced by over a million animals during the first four months of Covid confinement due to the reduction of vehicles on US roadways (Rott, 2020).

To make these types of changes permanent, however, requires activism. There are signs this may be more likely in the post-Covid era. Many businesses, for example, see enormous cost benefits in reducing office space and supporting remote work from home. Also, the low cost of petroleum has encouraged investors to seek out potentially more profitable renewable sources of energy (Penn, 2020). Rationalization will at times coincide with the shift towards greater sustainability, but that is not usually the case. To assure we remain on the right track will require sets of rules and regulations that favor conservation of resources and energy over increased production, and reflect a greater regard for human dignity and ecological equilibrium than profit maximization.

Place social equity at the core of plans for a sustainable future

There was a time not long ago when the predominant approach to thinking about sustainability ignored social inequality and relied primarily on economic growth as the panacea for the world's environmental problems. The environmental Kuznets Curve was an exemplar in this regard. It predicted that as economies grew and people earned higher income they would, (a) have the freedom to be concerned about the environment (as opposed to other more pressing economic concerns), and (b) be willing to do something about it. That, unfortunately, is not how things usually work out. In general, when people earn higher income they eat more meat, buy more cars, live in bigger houses, and not unexpectedly, wreak havoc on the environment through their increased consumption.

At both the national and global scale, people of lesser means have significantly smaller ecological footprints than those at the top of the income scale. A 2015 Oxfam study found the poorest half of the world population contributes only 10 percent to the world's individual-sourced carbon emissions (Gore, 2015). Meanwhile, about 50 percent of these emissions can be attributed to the wealthiest 10 percent. On a relational scale, inequality creates crises of comparison wherein people of lesser means feel compelled to spend beyond their earnings – i.e., get into debt – just to keep up with the perceived norms set by those with greater visibility and higher status. This is especially evident when comparing across high income countries: Both the quantity of municipal waste collected and the consumption of beef, for example, are twice as high in the United States – a highly unequal society – as they are in South Korea where the disparity between rich and poor is much less (Dorling, 2017).

Focusing on the other side of the wealth scale, environmental justice research shows the disproportionate impact environmental hazards have on the not-rich (Bullard, Mohai, Saha, & Wright, 2007). Be it communities proximate to toxic industries, agricultural workers or nations displaced by rising sea levels, these are the kinds of groups most directly impacted by the excesses of industrial capitalism and, not surprisingly, the ones most concerned about environmental risks.

Though the Kuznets Curve has fallen out of favor as a way of anticipating our ecological future, there is little in the way of an agreed upon plan that would take into serious consideration inequality as a key determinant of sustainability going forward. In the interdisciplinary anthology, *Just Sustainabilities*, Duncan McClaren (2003) proposes one possibility in this vein; an "environmental space" model which takes into consideration both the Earth's sink limits for pollution and the disproportionality of consumption in the world's population. The ideal goal here would be to set strict limits on the quantity of carbon dioxide, for example, that could be released into the atmosphere, tie this back to fossil fuel consumption, and then limit national-level consumption on a per capita basis. In this calculus, prosperous countries that consume the most would need to make the greatest reductions while many poorer countries would not have to make any reductions at all.

This effectively turns on its head the population theories of Thomas Malthus who, in the late 18th century, proposed that it was poor people in highly populated regions who presented the greatest threat to civilization. Whether it be Malthus' population theory or the Kuznets Curve, the primary concern among these influential models has been keeping at bay potential threats to the lifestyles and needs of the world's most privileged inhabitants.

From a socioecological perspective, however, inequality is the problem. A more just global society where the socioecological costs and benefits of natural resource conservation are equitably distributed has to be an integral part of the solution.

At the personal level, thinking about how your own consumption compares with that of others with fewer resources than yourself is a good place to begin making changes in your personal ecological impact and its implications for our collective future. Though a greater challenge, selecting sustainability-minded leaders and pushing lawmakers towards the implementation of structural changes that would lessen social inequality would make an even bigger difference.

Use intersectionality to our collective advantage

The United States is a country of remarkably diverse origins. As a national personality trait, this is both our greatest strength and weakness. We share a history, after all, stained with the violence of forced enslavement and the genocide of indigenous people. Racism and xenophobia have also long been deployed to dehumanize waves of immigrants who, though essential for our economy, are treated as second-class citizens, unworthy of basic rights, and at times, a living wage. Even for those not marked as racially or non-normatively different, prejudice has worked to divide Americans from each other, blinding them to their common interests as workers and citizens in a free, democratic society.

And yet, it is our diversity and verve to stand up for the ideals of an egalitarian and free society laid out in our founding documents – penned largely by slave-owners, granted – that make us unique and inspire others. One need look no further than the election of Barack Obama, the first African American president, in 2008 to see both how far the United States has come in trying to attain its ideals, and how much it has left to go. As the son of a white American mother and a father from Kenya – both graduate students at the University of Hawaii when they met – Obama, was also the first intersectional president, reminding us of the overlapping categories of difference most of us possess.

We of course can selectively choose these categories to delineate the metaphorical borders between us, such as when families suppress the Black or indigenous roots of their genealogy so as to emphasize whiteness, for example. If we choose to recognize our own intersectionality, however, we can break down barriers. This is evident in the gradual though steady shift in attitudes towards gay marriage and rights granted to the LGTBQ population. It is difficult to sustain prejudicial attitudes towards an entire class of people when they are open about their identity and you must reconcile the fact that they are often literally your brothers and sisters.

In a similar way, Black Lives Matter is a call for intersectional recognition in the face of structural racism; a reminder that we are indispensable to each other. From the country's origins where labor and land were forcibly extracted from non-Europeans, to the Covid-19 pandemic where African Americans and Latinxers are overrepresented among legions of essential workers, the nation's economy has been built up and sustained upon an unspoken system of racial caste. BLM thus puts forwards a demand that systemic racism be dismantled, replaced by a society which treats all its members with dignity, and where categories of difference serve not as a source of derision and division, but of unity and collective strength.

Intersectionality also sheds light on how categories of difference have been used to buffer us from some of our most egregious acts of environmental harm. According to the United Nations, the largest consumer contribution to greenhouse gases is not transportation but meat consumption. Yet, how many of us hunt for our own food? Who is slaughtering the animals and processing this primary source of protein for our consumption? In the United States, meatpacking, one of the least desirable occupations in the economy, is done primarily by immigrants. Which neighborhoods in American cities are most exposed to environmental hazard, including air pollution, lead poisoning, and the legacy of industrial contamination buried underground? Disproportionately, these are the poor and communities of color. And, where on the globe are people receiving the brunt of meteorological volatility, desertification, and displacement due to rising sea levels caused by global climate change? Again, this is occurring primarily among the non-white nations of the global south.

Categories of difference thus buffer our exposure to anthropogenic impacts on the environment. However, if we chose, we can use these differences as a starting point for working towards a just, equitable, and sustainable society. Intersectionality invites us to make these connections, by both bringing into relief long histories of violence and exploitation too often ignored, and reminding us that, if we all occupy more than one category at a time, we also possess the potential for common cause, empathy, and mutual understanding. On the personal level, this may not come easy. It involves placing yourself outside your comfort zone; making a real effort to understand people who grew up in different circumstances from yourself. If you are serious about this – and here is an area most native-born Americans could put some effort – learn to speak a second language. Language, of course, is not just a collection of words and grammar put together in just the right way; more importantly, it is a conduit for empathy. Fortunately, a growing diversity across the United States – about 15 percent of US residents are foreign-born – provides ripe opportunities for practicing and picking up languages outside the classroom.

Understand the transformative nexus of social structure and agency

Sociologist Eric Klinenberg has written elegantly about the role of physical structures in modern life that directly affect the way we interact with one another in his book, *Palaces for the People* (2018). Places like the public library, integrated public pools, and union halls have played essential roles in getting people across categories of difference to interact with each other. Unfortunately, legacies of racism, sexism, and classism have created major obstacles in keeping these institutions alive. In the case of public pools, these tended to be relatively integrated social spaces until they became integrated by sex, meaning Black men and white women might find themselves in the same social space. In white supremacist culture, this was intolerable and, in the 1960s and 1970s, hundreds of public pools, instead of following Civil Rights legislation requiring integration, simply shut down.

If you believe in the power of diverse multiplex ties – and I hope you do after reading this book – you should work to restore and sustain institutions like public pools, community gardens, and libraries; essential elements of social structure that bridge the gap across categories of difference. Structures, though solid and seemingly immovable, are the results of deliberate effort on the part of many. They can be used to extract ever greater amounts of profit from the masses such as online platforms that flatter, reassure, and generally facilitate your ability to spend time and money on impulse purchases. Or, they can be used to generate trust and good will, by providing the context for sustained community interaction.

Social structure also reminds us that, though values are important, more often than not it is the physical space and objects that surround us that determine our behavior. For example, where I live in a 1950s-era suburb of Burlington, Vermont – a politically liberal community that prides itself in its progressive ecological ideals – there are currently three different garbage and recycling services that drive through the neighborhood twice a week to pick up refuse. Along with the noise pollution this produces, the cumulative amount of diesel exhaust wafting into people's yards and homes is noxious and unnecessary.

By contrast, in my wife's hometown of 8,000 inhabitants in southeastern France which leans politically to the right, door-to-door garbage pick-up does not even exist. Instead, every block or so of residents have their own trash drop-off site where, typically, every day before heading off to work or school you deposit your sorted trash and recycling in the appropriate bin. This is not unique, by the way, to small towns; in Paris and Marseille, you can find the same color-coded brown, yellow, green, and blue bins which, though perhaps less profitable than competing garbage services, are less polluting and, not incidentally, force you to walk outside in public where you may run into other people.

Without ever having to think of it, the structure of trash collection in France reduces your ecological footprint while, perhaps, making you slightly more social than you would be otherwise. This point is key because a lot of academic research aimed at encouraging sustainable behavior focuses on trying to change people's values and beliefs regarding the environment. The encouraging news is that, on the whole, people care about the environment. The bad news is what is known as the *attitude-behavior gap*. That is, in the aggregate, people's pro-environmental attitudes rarely result in meaningful behavioral change (Shove, 2010). Most Americans know climate change presents an imminent threat to life as we know it, and yet our consumption of fossil fuels continues to rise.

To put it another way, many social scientists have placed too much faith in human agency and not enough thought into changing social structure. Instead of trying to shape people's minds so as to behave in a more sustainable fashion, we should pay more attention to the nexus of structure and agency, an area where social change may be able to rectify ecological insult. For sociologist Elizabeth Shove, this means bringing a critical eye to practice in consumer society (Shove, 2003).

Electric clothes dryers, for example, are a convenient yet wasteful use of energy. In the United States, 92 percent of households use electric dryers (Wachunas, 2019). By contrast, in both Italy and China only 3 percent of households own this appliance, opting instead to hang their clothes out to dry on the line. The difference between these two outcomes likely has little to do with ecological values. Instead, relatively cheap electricity has made the use of electric clothes dryers the default practice in the United States. Change the structure and you can change the practice, regardless of professed individual-level or corporate values.

The nice part of this approach is that it works for assessing areas in your own life, as well as where you would like to make improvements in terms of broader social change. Think of all the things you do not really need; now get rid of them. Along with living without a dryer (many before have done so and survived), could you make do without a car? Why not try it out? If religious metaphors help, when temptation is out of reach you are unlikely to commit the sin. Brushing your teeth, tying your shoes, wearing a mask during a global pandemic – these are all practices that work best when you do not have to think about them.

Once you have structured your life in a way that is sustainable, equitable, and just, don't think, don't philosophize, just practice, practice, practice! The more you do so in collaboration with others, and within physical, institutional, and policy structures that support it, the deeper and longer lasting the change will be you want to bring about. And that, dear reader, is how sociology will save the planet, though it is up to all of us collectively to do the work. Let's get started.

Questions

1   How does what you have learned about sociology change the way you think about environmental problems? If some asked you, "What is environmental sociology?", what would you have for a response?

2   How would you be able to utilize each of the five socioecological strategies outlined in this chapter in your own life to think about and act to bring about positive socioecological change?

References

Bullard, R. D., Mohai, P., Saha, R., & Wright, B. (2007). *Toxic wastes and race at twenty 1987–2007: Grassroots struggles to dismantle environmental racism in the United States*: United Church of Christ Justice and Witness Ministries.

Cagle, S. (2019). Bees, not refugees': The environmentalist roots of anti-immigrant bigotry. *The Guardian*, August 16. Retrieved from https://www.theguardian.com/environment/2019/aug/15/anti. Accessed September 17, 2021.

Dorling, D. (2017). Is inequality bad for the environment? The Guardian, July 4. Retrieved from https://www.theguardian.com/inequality/2017/jul/04/is-inequality-bad-for-the-environment. Accessed September 17, 2021.

Eaton, C., & Matthews, C. (2020). Dakota access pipeline must shut down, judge Rules a federal court ruled the pipeline, which has been carrying oil since 2017, must stop until a key environmental approval is redone. *Wall Street Journal*, August 6. Retrieved from https://www.wsj.com/articles/dakota-access-pipeline-must-shut-down-judge-rules-11594052464. Accessed September 17, 2021..

Gilio-Whitaker, D. (2019). *As long as grass grows: The indigenous fight for environmental justice, from colonization to Standing Rock*: Beacon Press.

Gore, T. (2015). *Extreme carbon inequality: Why the Paris climate deal must put the poorest, lowest emitting and most vulnerable people first*: Oxfam International. Retrieved from http://policy-practice.oxfam.org.uk/publications/extreme-carbon-inequality-why-the-paris-climate-deal-must-put-the-poorest-lowes-582545. Accessed September 17, 2021.

Gould, S. J., & Eldredge, N. (1977). Punctuated equilibria: The tempo and mode of evolution reconsidered. *Paleobiology, 3*(2), 115–151.

Harrison, J. L. (2019). *From the inside out: The fight for environmental justice within government agencies*: MIT Press.

Head, L., Klocker, N., & Aguirre-Bielschowsky, I. (2019). Environmental values, knowledge and behaviour: Contributions of an emergent literature on the role of ethnicity and migration. *Progress in Human Geography, 43*(3), 397–415.

Klinenberg, E. (2018). *Palaces for the people: How social infrastructure can help fight inequality, polarization, and the decline of civic life*: Broadway Books.

Krymkowski, D. H., Manning, R. E., & Valliere, W. A. (2014). Race, ethnicity, and visitation to national parks in the United States: Tests of the marginality, discrimination, and subculture hypotheses with national-level survey data. *Journal of Outdoor Recreation and Tourism, 7*, 35–43.

Macias, T. (2016a). Ecological assimilation: Race, ethnicity, and the inverted gap of environmental concern. *Society & Natural Resources, 29*(1), 3–19.

Macias, T. (2016b). Environmental risk perception among race and ethnic groups in the United States. *Ethnicities, 16*(1), 111–129.

Mascarenhas, M., Carrera, J., Richter, L., & Wilder, E. (2017). Diversity in sociology and environmental sociology: What we know about our discipline. *ASA ETS Section News*, 1–5.

McLaren, D. (2003). Environmental space, equity and the ecological debt. In J. Agyeman, R.D. Bullard & B. Evans (Eds.), *Just sustainabilities: Development in an unequal world*: MIT Press, 19–37.

McPherson, M., Smith-Lovin, L., & Brashears, M. E. (2006). Social isolation in America: Changes in core discussion networks over two decades. *American Sociological Review, 71*(3), 353–375. doi:10.1177/000312240607100301

McPherson, M., Smith-Lovin, L., & Cook, J. M. (2001). Birds of a feather: Homophily in social networks. *Annual Review of Sociology, 27*(1), 415–444.

Melley, B. (2020, July 22, 2020). Sierra Club apologizes for founder John Muir's racist views. *Washington Post.*

Penn, I. (2020). Oil companies are collapsing, but wind and solar energy keep growing. *The New York Times, 7.*

Purdy, J. (2015). Environmentalism's racist history. *The New Yorker, 13.*

Rott, N. (2020). Animals see a silver lining during pandemic as fewer are killed on the road. *NPR.*

Shove, E. (2003). *Comfort, Cleanliness and Convenience: The Social Organization of Normality*: Berg.

Shove, E. (2010). Beyond the ABC: climate change policy and theories of social change. *Environment and Planning A, 42*(6), 1273–1285.

Taylor, D. E. (2014a). *The state of diversity in environmental organizations*: University of Michigan.

Taylor, D. E. (2014b). *Toxic communities: Environmental racism, industrial pollution, and residential mobility*: New York University Press.

Wachunas, J. (2019). Hang drying revolution. *Medium*, August 4. Retrieved from https://info-25289.medium.com/hang-drying-revolution-3d189d01ab22. Accessed September 17, 2021.

Index